English for Specific Purposes

Edited by Thomas Orr

Case Studies in TESOL Practice Series

Jill Burton, Series Editor

Teachers of English to Speakers of Other Languages, Inc.

Typeset in Berkeley and Belwe
by Capitol Communication Systems, Inc., Crofton, Maryland USA
Printed by Kirby Lithographic Company, Arlington, Virginia USA
Indexed by Coughlin Indexing Services, Annapolis, Maryland USA

Teachers of English to Speakers of Other Languages, Inc.
700 South Washington Street, Suite 200
Alexandria, Virginia 22314 USA
Tel. 703-836-0774 • Fax 703-836-6447 • E-mail tesol@tesol.org • http://www.tesol.org/

Director of Communications and Marketing: Helen Kornblum
Managing Editor: Marilyn Kupetz
Copy Editor: Ellen Garshick
Cover Design: Capitol Communication Systems, Inc.

ISBN 0-939791-95-1
Library of Congress Control No. 01-135521

Dedication

To Sakurako, my wife and best friend, with love.

Table of Contents

Acknowledgments

I would like to express my deepest appreciation to Jill Burton, Mark Freiermuth, Ellen Garshick, and Marilyn Kupetz for their kind assistance with this volume, as well as to the authors, whose hard work and infinite patience made this book possible.

Series Editor's Preface

The Case Studies in TESOL Practice series offers innovative and effective examples of practice from the point of view of the practitioner. The series brings together from around the world communities of practitioners who have reflected and written on particular aspects of their teaching. Each volume in the series will cover one specialized teaching focus.

◈ CASE STUDIES

Why a TESOL series focusing on case studies of teaching practice?

Much has been written about case studies and where they fit in a mainstream research tradition (e.g., Nunan, 1992; Stake, 1995; Yin, 1994). Perhaps more importantly, case studies also constitute a public recognition of the value of teachers' reflection on their practice and constitute a new form of teacher research—or teacher valuing. Case studies support teachers in valuing the uniqueness of their classes, learning from them, and showing how their experience and knowledge can be made accessible to other practitioners in simple but disciplined ways. They are particularly suited to practitioners who want to understand and solve teaching problems in their own contexts.

These case studies are written by practitioners who are able to portray real experience by providing detailed descriptions of teaching practice. These qualities invest the cases with teacher credibility, and make them convincing and professionally interesting. The cases also represent multiple views and offer immediate solutions, thus providing perspective on the issues and examples of useful approaches. Informative by nature, they can provide an initial database for further, sustained research. Accessible to wider audiences than many traditional research reports, however, case studies have democratic appeal.

◈ HOW THIS SERIES CAN BE USED

The case studies lend themselves to pre- and in-service teacher education. Because the context of each case is described in detail, it is easy for readers to compare the cases with and evaluate them against their own circumstances. To respond to the wide range of language environments in which TESOL functions, cases have been selected from EFL, ESL, and bilingual education settings around the world.

The 12 or so case studies in each volume are easy to follow. Teacher writers describe their teaching context and analyze its distinctive features: the particular demands of their context, the issues they have encountered, how they have effectively addressed the issues, what they have learned. Each case study also offers readers practical suggestions—developed from teaching experience—to adapt and apply to their own teaching.

Already in published or in preparation are volumes on

- academic writing programs
- action research
- assessment practices
- bilingual education
- community partnerships
- content-based language instruction
- distance learning
- gender and TESOL
- grammar teaching in teacher education
- intensive English programs
- interaction and language learning
- international teaching assistants
- journal writing
- literature in language teaching and learning
- mainstreaming
- teacher education
- technology in the classroom
- teaching English as a foreign language in primary schools
- teaching English from a global perspective
- teaching English to the world

◈ THIS VOLUME

The range of contexts and content of the chapters in this volume truly illustrates that although effective English teaching is always purposeful, always specific, language use in a specific setting also always has general applications. Whatever one's teaching context, there is much to learn about curriculum and course planning, implementation, and evaluation from these case studies of TESOL in industrial, domestic, business, sport, tertiary, and professional settings.

Jill Burton
University of South Australia, Adelaide

INTRODUCTION

The Nature of English for Specific Purposes

Thomas Orr

English for specific purposes (ESP) is an exciting movement in English language education that is opening up rich opportunities for English teachers and researchers in new professional domains. The growing demand for highly proficient speakers of specialized academic and workplace English is drawing increasingly large numbers of teachers into the ESP profession and awarding them higher salaries and prestige than were previously given to language instructors. Those who are interested in learning more about these developments, as well as those who would like to prepare for employment in this field, will find this volume in TESOL's Case Studies series an ideal place to begin.

◈ THE BASICS

ESP currently possesses three specific referents in the world of English language education:

1. specific subsets of the English language that are required to carry out specific tasks for specific purposes

2. a branch of language education that studies and teaches subsets of English to assist learners in successfully carrying out specific tasks for specific purposes

3. a movement that has popularized the ESP profession and its work with ESP discourse

The ESP that is primarily taught or researched consists of spoken and written discourse in academic and workplace settings, which is unfamiliar to most native and nonnative speakers and thus requires special training. Specific-purpose English includes not only knowledge of a specific part of the English language but also competency in the skills required to use this language, as well as sufficient understanding of the contexts within which it is situated. Although the name can be misleading, ESP does not refer to English or English language education for any specific purpose. All education exists for specific purposes, but only English education for highly specialized purposes, which require training beyond that normally received in Grades K–12 or the ESL/EFL classroom, interests ESP professionals.

English that is commonly known by the average native or nonnative speaker is called *English for general purposes* (EGP). General-purpose English comprises the common core of English that is shared by most of its speakers. Learning general-purpose English typically begins at home for native speakers and in the ESL/EFL classroom for nonnative speakers. When taught, EGP is presented as a linguistic system to a wide range of learners for application in the most general of potential circumstances, whereas ESP is taught as a tailor-made language package to specific communities of learners with highly specialized language needs. A simple comparison of some general and specific purposes that require general-purpose English and specific-purpose English illustrates the distinction between these two domains more clearly (see Table 1). Although the English needed for specific academic and career purposes frequently contains large portions of general-purpose English, the nature of these specialized purposes and the tasks that are necessary to achieve them may remain unusual enough that they require special training from qualified experts who understand the context and can provide appropriate instruction. This is the job of professionals in ESP.

❖ THE CASE STUDIES

The chapters in this volume exemplify research and teaching activities that fall within the ESP profession in university and workplace settings. They illustrate the nature of ESP as currently understood and practiced in a variety of different contexts and locations, with all of its admirable strengths and a few of its yet-to-be-resolved weaknesses.

The case studies in Part 1 exemplify ESP in university settings. Feak and Reinhart (chapter 1) introduce a program at the University of Michigan for students preparing to enter U.S. law schools; Hussin (chapter 2) details a program for nursing students at Flinders University in Australia; and Boyd (chapter 3) describes a program at Columbia University in the United States for students and working

TABLE 1. COMPARISON OF GENERAL AND SPECIFIC PURPOSES

General English Purposes	Specific English Purposes
To initiate conversation with a stranger	To negotiate a merger
To make a doctor's appointment	To produce software documentation
To order food at a restaurant	To engage in courtroom debate
To read a local newspaper	To announce an aircraft's position to the control tower
To report a crime to the police	To understand pesticide application instructions
To fill out a credit card application	To complete a grant proposal
To comprehend a TV news program	To read technical specifications
To address an envelope	To explain how to operate a crane
To shop via the Internet	To make a stock trade on the trading floor
To exchange letters with a friend	To write a medical prescription

professionals in business. Magennis (chapter 4) continues the academic lineup with details on a program in Portugal at the Instituto Superior de Assistentes e Interpretes for students in tourism; López Torres and Perea Barberá (chapter 5) describe ESP work in Spain for students and shipbuilders at the University of Cádiz; and Papajohn, Alsberg, Bair, and Willenborg (chapter 6) conclude Part 1 with an outline of ESP training for international teaching assistants at the University of Illinois, in the United States.

In Part 2, on ESP in the workplace, Eggly (chapter 7) outlines a program for international medical graduates at a Detroit hospital, in the United States; Baxter, Boswood, and Peirson-Smith (chapter 8) describe an ESP consultancy for the horse-racing industry in Hong Kong; and Gordon (chapter 9) illustrates the work of ESP in a typical U.S. factory setting. In chapter 10, Garcia provides another perspective on ESP in industry with an overview of work at 25 factories in the Chicago area, in the United States; Orsi and Orsi (chapter 11) report on a special English training program they designed to prepare brewers in Argentina for professional training in Europe; and Noden (chapter 12) concludes the volume with details on a consultancy for a large home-cleaning service in the suburbs of Washington, DC, in the United States. All 12 chapters not only illustrate the fine work that has been done so far in ESP but also provide ideas and inspiration for the next generation of ESP programs that will evolve from these models.

Where will the ESP movement take us next? If systematic attention to actual needs continues to be its hallmark, ESP will clearly advance further in its study of specialized English discourse and in its development of effective methodologies to teach it. Fiercer competition in academe and the workplace will heighten the demand for competent speakers of specialized English, and ESP professionals who can teach this discourse will be needed in the future more than ever before. The fine programs featured in the chapters that follow are an excellent starting place for understanding the nature of ESP and suggest where fruitful professional activity might be found in the future.

◈ CONTRIBUTOR

Thomas Orr is professor in the Center for Language Research at the University of Aizu, in Japan, where he researches technical discourse in the fields of science, engineering, and information technology. He is founder of the Japan Conference on English for Specific Purposes, former chair of TESOL's ESP Interest Section, and associate editor for ESP submissions to *Transactions on Professional Communication,* published by the Institute of Electrical and Electronics Engineers (IEEE). He is also a founding board member and chair of the Professional English Research Consortium, advisor to the Institute for Professional English Communication, and content consultant for the Test of Professional English Communication. His work has been published by IEEE, Wiley-InterScience, TESOL, Shogakukan, and others.

PART 1

ESP for Language Learners in the University

CHAPTER 1

An ESP Program for Students of Law

Christine Feak and Susan Reinhart

◈ INTRODUCTION

Presessional programs can be particularly valuable to nonnative English speakers entering rigorous graduate programs, such as those leading to a master of business or law, because the curricula of such programs tend to be regimented, leaving little opportunity to take supplemental language courses during the regular academic year. Though presessional English for academic purposes (EAP) programs and business English programs are well established throughout the United States, very few programs address the special needs of incoming law students. At present, the most institutions can do for students who are interested in presessional language study is to funnel them into existing EAP programs, most of which, in part because of their broad scope, are ill-equipped to address the very specific linguistic and cultural demands of law school.

In this chapter, we describe an ESP program developed at the University of Michigan, in the United States, for nonnative speakers of English who have been accepted into a competitive U.S. law school, usually in the master of law (LLM) program. We discuss the evolving framework and curriculum of the program, considerations in designing the program, the utilization of discoursal research, and directions for future research, and we offer advice for the implementation of similar programs elsewhere.

◈ CONTEXT

The 6½-week English for Legal Studies (ELS) program is offered each summer alongside the English for Business Studies program and the EAP program at the English Language Institute of the University of Michigan. The youngest and the smallest of the three programs, ELS was begun in 1995.

Before 1995, law students interested in attending or required to attend a presessional language program at the university were placed into the EAP program. The number of law students in the EAP program was small (on average, 2–4 of 30 students); nevertheless, each year at least some law students in the EAP program were there to satisfy law school language requirements. These students were attractive to the law school but were potentially risky admissions because of their low scores on the Test of English as a Foreign Language (TOEFL)—meaning 570–600 on the paper-based test (230–253 on the computer-based test). A follow-up investigation on these law students, however, revealed that the EAP program was not adequately preparing them for their law programs, particularly in terms of reading and writing. Despite their high level of proficiency in English and their better-than-average success in the EAP program, students reported having difficulty managing legal case reading, writing seminar papers, and participating in seminar classes.

The Study of Law in the United States

The nature of the difficulties faced by our students and, most likely, by other nonnative-English-speaking law students becomes clearer with a general understanding of the study of law in the United States. Law school in the United States is a graduate program that generally requires 3 years of full-time study. Successful completion leads to a doctor of jurisprudence (JD) degree. JD students take approximately 85 credits of course work, including required core courses, electives, a writing and research class, and a senior research seminar. The LLM program, which nearly all of our students take, differs from the JD in length and curricular flexibility. It consists primarily of courses and seminars, which JD students may also attend, and affords students an opportunity to pursue individual research. Students in this program may freely select courses and seminars according to their interests, totaling 24 credits over the academic year. To obtain a degree, however, LLM students must generally complete a lengthy research paper on a topic of their choice as part of a seminar or a supervised independent research project.

Whereas British law school classes use textbooks that are "similar in format to . . . textbooks in other disciplines" (Howe, 1990, p. 8), U.S. law school courses other than legal writing or research classes generally do not use textbooks. For each core course, a law student has one basic text, called a *casebook*, which is a primary teaching tool "containing texts of leading appellate court decisions that in common law tradition often serve as precedent in a particular field of law, such as contracts or torts, sometimes together with commentary and other features that might be useful for class discussion and further understanding of a subject" (Black, 1991, p. 148).

A typical casebook contains an introduction to a short series of cases, the cases themselves, and a series of notes and questions for the cases. The cases are often abridged so as to highlight one or several points of law. In some courses, such as those in criminal or tax law, the casebook is accompanied by a book of statutes or regulations. Unlike most U.S. university textbooks, which highlight important information and suggest what students should learn, a casebook does not indicate what exactly is to be learned. Students must figure out on their own the main point or issue illuminated by the case and the aspects of the case relevant to the issue. Through this process, students develop cognitive skills and strategies for dealing with cases in ways that experienced legal professionals do—one of the key objectives of law school. In addition, after reading and understanding the cases, students are expected to have acquired knowledge of an entire area of law.

Thus, in law schools, students spend the vast majority of their time reading cases. During a year-long course, for example, it is not uncommon for students to read 125 cases ranging in length from 2 to 20 pages, along with a series of notecases, which are summaries of related cases. Although the amount of reading may actually be less than in other disciplines, for new students the time required to read even the shortest case is substantial because of the newness of the content and genre, the density of information, and the frequent use of common words with uncommon meanings.

Generally, law professors in the United States use a modified Socratic method in the classroom and devote much less time to lectures than professors in other disciplines do. Consequently, students may be randomly or systematically called on in class, expected to brief (i.e., discuss and analyze) a case or part of a case, and sometimes asked to defend a position or critique the court's reasoning. Students are expected to actively participate in class from the start.

First-year classes are large compared with most graduate school classes (sometimes having as many as 100 students), so a student may be called on only a few times during a semester. Because students generally do not know when they will be called on, they must always be prepared to speak when asked. Better students, however, try to volunteer often because they want to appear eager and establish their presence among the others. Seminars, consisting of 15–25 students, place even greater pressure on students to participate, along with an additional burden of presenting personal research. Seminars also often require students to produce a major research paper following academic legal writing conventions, which students are assumed to know. Thus, seminars can be linguistically challenging for nonnative English speakers.

The ESP Program

Given this understanding of the linguistic demands faced by international students in an LLM program, the ELI decided to design an ESP course specifically for law, a goal we felt could be realized because one of our regular EAP instructors had just completed a law degree. Before fully developing our ESP program, we first offered a pilot program in 1994 as part of the EAP program, focusing mainly on case reading for a small group of law students. The response to this trial was overwhelmingly positive. In their evaluations, the students said that they wanted more time to develop their case-reading skills and more instruction in writing. Thus, we determined that developing a more specialized curriculum for law would be both feasible and desirable.

One of our first activities before designing the ESP program was to search for other ESP programs in the United States that focused on law. Unsurprisingly, we found very few programs that focused on legal English and none that centered only on English for academic legal purposes (EALP). For the programs offering course work in legal English, the teaching of EALP seemed to be secondary to other goals, such as teaching general EAP or providing an introduction to U.S. law, often to students ranging from legal assistants and secretaries to lawyers. Thus, the need for an intensive program devoted entirely to EALP seemed to extend beyond our own university.

❖ DESCRIPTION

The program provides approximately 18 hours of instruction each week in the form of classes, movies, individual appointments, a workshop, and law-related field trips. Enrollment in the program is limited to 15 students per year, a limit imposed to allow students to receive individual attention from the instructors and to promote interaction among the participants.

Admissions

Because our institute is a service unit at the University of Michigan, all students admitted to the university's law school are guaranteed admission to our program. Non–University of Michigan students are admitted if they have the requisite TOEFL score and will attend an LLM program at a comparably ranked law school. This requirement exists because the law school stresses that, for the program to receive its support, all students should be more or less similar in terms of legal background and language ability. Generally, the number of applicants to the program matches the number of places available. However, should the number of qualified applicants increase beyond the program's current capacity, a second class could be added.

Student Profile

The ESP students come from many different countries and represent a broad spectrum of legal, cultural, and political viewpoints, which greatly enriches the knowledge and widens the perspectives of everyone. Over the years, Asian students (from Japan, Korea, Taiwan, and Thailand) have constituted the largest group of students (40%), and the remaining 60% have come from such diverse countries as Argentina, Armenia, Azerbaijan, Belarus, Brazil, Chad, Chile, Costa Rica, Ethiopia, Italy, Mexico, Peru, and Russia. Students in the program have all graduated from an undergraduate law program and have worked an average of $4^1/_2$ years in the field. Thus, they come into the program with strong legal backgrounds. These legal professionals have been employed in various capacities, including attorney in a law firm, corporate lawyer, public prosecutor, judge, professor of law, and lawyer in a government ministry. They have come to the United States to further their legal education, usually in an LLM program but sometimes to pursue research as visiting scholars. Although the average TOEFL score of the students is high (604 on the paper-based or 250 on the computer-based test), they have entered our program because their funding agency, the law school they will attend, or they themselves are concerned about their ability to meet the linguistic challenges of law school.

To learn more about our students' language abilities, before the program begins we ask them to complete a questionnaire, part of which is a language skills inventory or self-assessment. Table 1 shows how students have responded since 1995 to questions about their weakest and strongest skills in English. As is often the case with international students, our group was quick to point out perceived deficits in language ability, with roughly equal numbers citing speaking, listening, and writing as their weakest skills. There was more agreement, however, in evaluating their strongest skill, which slightly more than two thirds believed was reading. Although we agree that many students have difficulties with speaking, listening, and writing, we would also argue that, contrary to their self-assessment, reading is in fact one of their weakest skills, given their lack of experience in reading U.S. legal cases and the demanding nature of such reading, which we describe below.

TABLE 1. ELS STUDENTS' ASSESSMENT OF THEIR LANGUAGE SKILLS, 1995–1998 (%, N = 41)

Skill	Weakest	Strongest
Speaking	33	10
Listening	33	10
Reading	0	69
Writing	33	10
No skill	0	1
All skills	1	0

Program Objectives

The general objective of the ESP program is to prepare international students during the summer for law school in the fall. Its primary aim is to familiarize students with law school culture along with the language and academic skills needed to succeed in a rigorous LLM program.

Working from our general understanding of law school culture, described in the Context section; interviews with recent law school graduates and former students; evaluations of law school texts; and the analyses of homework assignments, examinations, and videotapes of law school classes, we identified eight competencies important to our program, some of which have been identified by Harris (1992) as well:

1. ability to handle cases, including briefing (both orally and in writing) as well as applying legal principles to the facts of a case

2. knowledge of the U.S. court system and understanding of common law and precedent

3. familiarity with legal resources (e.g., a law library) and finding legal material

4. use of electronic legal databases (e.g., Westlaw and LexisNexis)

5. ability to write academic legal documents (e.g., memos, briefs, and syntheses) and understand legal footnoting procedures

6. ability to carry out and present legal research

7. proficiency in comprehending and taking part in legal dialogues

8. awareness of classroom expectations, including class participation and exam taking

Using these competencies as a guide, we designed four interrelated courses (Processing Legal Materials, Academic Legal Writing, Interactive Listening and Speaking, and Researching Legal Issues) and an 8-hour workshop (Languages of the Law) for our annual summer ESP program.

Classes

Processing Legal Materials

Processing Legal Materials meets 5 hours per week and covers three main areas: case reading and discussion, legal terms and legal processes, and exam taking. The main goal of this class is to help students learn strategies for reading legal texts, primarily cases that deal with U.S. common or judge-made law, along with some cases in statutory and constitutional law. The cases come from common areas of law: contracts, torts, criminal law, and patent law. Although legal cases pose the greatest challenge for us in terms of teaching, they are essential to an EALP program because cases are the foundation of most law school classes. Given this emphasis on cases, this course, unlike the others in our program, has always been taught by an EAP instructor who also is a lawyer and can therefore make expert reading strategies explicit.

General academic reading strategies do not necessarily translate into good case reading strategies because of the distinctive characteristics of a case or court opinion, such as its structure. Our first task, then, involves helping students become aware of the parts of a case and understand the ways in which these various parts are related. Generally a case contains the following elements:

- the relevant facts
- the history of the case in the court system (procedural facts)
- the legal issue
- the rules or legal principles
- the holding or decision of the court
- the reasoning (rationale) of the court
- the dissent

Students also learn to extract other information about a case, such as the publisher, the court, the court date, the judge who wrote the opinion, and the vote of the court. Such contextual information is often relevant to a full understanding of a particular case.

In addition to learning to read individual cases, students become aware of the importance of making connections among a series of related cases in a particular area of law, because it is a series of cases written over time that shape a legal principle. In view of this, another goal of the course is to develop strategies that are useful for finding common legal threads among cases, drawing conclusions, and then anticipating how the threads may continue. Finally, students learn strategies for critiquing a court's opinion and for applying a court's analysis to hypothetical situations.

In designing materials for this class that help foster expert reading skills, we drew on Lundeberg's (1987) study contrasting novice and expert legal readers. Table 2 provides a partial summary of the key differences instrumental in our materials development. Although to a novice reader the law may at first appear to consist of a series of rules that can be mechanically applied to a given situation, it is in fact a creative analytical process carried out by the court. Thus, memorizing definitions and statutes will not lead to an understanding of any given law. To understand a law, students must first understand the process by which the court

TABLE 2. DIFFERENCES BETWEEN EXPERT AND NOVICE READERS OF LEGAL CASES
(BASED ON LUNDEBERG, 1987)

Experts	Novices
1. are knowledgeable about the text type and thus the structure and the analytical strategies in reading a case.	1. are not knowledgeable about the text type.
2. go to the end of the opinion to find the holding (decision) of the court because this guides their reading of a case.	2. generally do not find or know the holding until after they have read the entire case.
3. synthesize information as they read a case—they merge the relevant facts, issues, rule and rationale into a cohesive whole.	3. "tend to focus more narrowly on one element of the case, demonstrating less connectedness in their discussion of the case" (p. 414).
4. show a sophisticated understanding of the creative process of judicial decision making. They generally evaluate the judge's decision.	4. often assume that legal decisions are somewhat mechanical and thus evaluation of the judge's decision is unnecessary.
5. make themselves aware of the court and date.	5. tend to ignore the court and date.
6. infer hypothetical fact patterns and their possible legal outcomes.	6. do not infer hypothetical fact patterns.

applies existing principles from other relevant cases to the facts in the current case in order to arrive at a reasoned decision. In other words, the students learn to understand the thinking of the court—the cognitive processes that underlie its decision.

Academic Legal Writing

Before entering our ESP program, most of our students have done little writing in English either for professional or for academic purposes. They are generally unfamiliar with legal academic genres and subgenres as well as with the general characteristics of academic writing. Thus, the first of the program's writing courses, Academic Legal Writing, has several goals:

- help students gain experience writing in English
- familiarize students with the basics of good academic writing in general
- provide students practice with typical academic legal writing tasks, such as preparing a written brief of a case, synthesizing multiple cases, and drafting an office memorandum

This class meets 5 hours per week, with additional time for individual appointments to discuss written work. The first two writing tasks in the class, briefing and synthesizing, give students additional guidance in understanding the structure of legal cases. This instruction overlaps with Processing Legal Materials because sorting out the various components of a case can be conceptually difficult. By working on both oral and written briefs and syntheses, we feel, the students gain a solid understanding of the cognitive and language skills they need to handle briefing tasks. Whereas briefing is primarily concerned with identifying the relevant parts of the case, synthesizing, in which students look carefully for similarities and differences in a series of cases, is one process by which students learn to deal with cases based on similar facts but resulting in different outcomes. Both briefing and synthesizing are important because they often create the foundation for the writing of more complex documents, such as memoranda and legal research papers.

We use the property case *Conti v. ASPCA* (1974; see Figure 1) to help students begin to identify the important elements of a case. This case works quite well as a first one because the court clearly states—uncharacteristically—the issues in the case (Paragraph 6) and its use of established principles of property law (Paragraphs 11–16) to arrive at its holding or decision. Moreover, the facts presented in Paragraphs 1–5 are straightforward. By working on the preparation of their briefs and syntheses through the use of sample texts, students become familiar with some of the characteristics of academic writing, such as hedging and qualification, complex relative clauses with prepositions, *-ing* clauses of result, and midposition adverbs.

The final task in Academic Legal Writing, writing a memorandum to a senior law partner, draws on the skills acquired in writing the brief and the synthesis. The memorandum is an objective, exploratory discussion of a hypothetical legal problem that requires students to read a number of cases and other documents relevant to the problem and extract from them the aspects that apply to the new problem. In other words, the students must "analyze the legal principles that govern the issues raised by the problem and apply those principles to the facts of a case" (Shapo, Walter, & Fajans, 1991, p. 69). In this task, the students evaluate the strengths and weaknesses of a potential case and recommend a course of action using the cases and documents they have read as support.

Researching Legal Issues

Researching Legal Issues meets 3 hours per week and requires 2 or more hours per week of research time. This course provides a general, hands-on introduction to the process of legal research, culminating in the production of a 7- to 10-page research paper on a topic of the students' own choosing. In this course, students receive professional training in the use of Westlaw, a computer-based legal research service that provides on-line access to a database of legal information. In addition, students become acquainted with the law library and its resources through library exercises, a series of research tasks that involve finding cases, locating treatises, and exploring special collections. The other main project in this class is helping students become familiar with the discourse conventions of published academic legal writing, which also apply to the writing of papers for law seminars. Having some familiarity with the characteristics of legal research paper writing can help give nonnative-English-speaking students an edge on their native-speaking counterparts at the beginning of a new term and throughout the law program, as research writing for seminars is rarely taught in law schools.

One of the main challenges in developing this class has been the paucity of existing materials. Although research paper writing is required of all JD students, most of the widely used legal writing textbooks focus only on professional writing in the workplace, devoting no attention to research papers. In addition, published

Conti v. ASPCA
77 Misc.2d 61, 353 N.Y.S.2d 288

MARTIN RODELL, Judge.

1 Chester is a parrot. He is fourteen inches tall, with a green coat, yellow head and an orange streak on his wings. Red splashes cover his left shoulder. Chester is a show parrot, used by the defendant ASPCA in various educational exhibitions presented to groups of children.

2 On June 28, 1973, during an exhibition in Kings Point, New York, Chester flew the coop and found refuge in the tallest tree he could find. For seven hours the defendant sought to retrieve Chester. Ladders proved to be too short. Offers of food were steadfastly ignored. With the approach of darkness, search efforts were discontinued. A return to the area on the next morning revealed that Chester was gone.

3 On July 5th, 1973 the plaintiff, who resides in Belle Harbor, Queens County, had occasion to see a green-hued parrot with a yellow head and red splashes seated in his backyard. His offer of food was eagerly accepted by the bird. This was repeated on three occasions each day for a period of two weeks. This display of human kindness was rewarded by the parrot's finally entering the plaintiff's home, where he was placed in a cage.

4 The next day, the plaintiff phoned the defendant ASPCA and requested advice as to the care of a parrot he had found. Thereupon the defendant sent two representatives to the plaintiff's home. Upon examination, they claimed that it was the missing parrot, Chester, and removed it from the plaintiff's home.

5 Upon refusal of the defendant ASPCA to return the bird, the plaintiff now brings this action in replevin [recovery of property].

6 The issues presented to the Court are twofold: One, is the parrot in question truly Chester, the missing bird? Two, if it is in fact Chester, who is entitled to its ownership?

7 The plaintiff presented witnesses who testified that a parrot similar to the one in question was seen in the neighborhood prior to July 5, 1973. He further contended that a parrot could not fly the distance between Kings Point and Belle Harbor in so short a period of time, and therefore the bird in question was not in fact Chester.

8 The representatives of the defendant ASPCA were categorical in their testimony that the parrot was indeed Chester, that he was unique because of his size, color and habits. They claimed that Chester said "hello" and could dangle by his legs. During the entire trial the Court had the parrot under close scrutiny, but at no time did it exhibit any of these characteristics. The Court called upon the parrot to indicate by name or other mannerism an affinity to either of the claimed owners. Alas, the parrot stood mute.

9 Upon all the credible evidence the Court does find as a fact that the parrot in question is indeed Chester and is the same parrot which escaped from the possession of the ASPCA on June 28, 1973.

10 The Court must now deal with the plaintiff's position, that the ownership of the defendant was a qualified one and upon the parrot's escape, ownership passed to the first individual who captured it and placed it under his control.

Continued on page 16

11 [1] The law is well settled that the true owner of lost property is entitled to the return thereof as against any person finding same. (In re Wright's Estate, 15 Misc.2d 225, 177 N.Y.S.2d 410) (36A C.J.S. Finding Lost Goods).

12 [2] This general rule is not applicable when the property lost is an animal. In such cases the Court must inquire as to whether the animal was domesticated or ferae naturae (wild).

13 [3] Where an animal is wild, its owner can only acquire a qualified right of property which is wholly lost when it escapes from its captor with no intention of returning.

14 Thus in Mullett v. Bradley, 24 Misc. 695, 53 N.Y.S. 781, an untrained and undomesticated sea lion escaped after being shipped from the West to the East Coast. The sea lion escaped and was again captured in a fishpond off the New Jersey Coast. The original owner sued the finder for its return. The court held that the sea lion was a wild animal (ferae naturae), and when it returned to its wild state, the original owner's property rights were extinguished.

15 In Amory v. Flyn, 10 Johns. (N.Y.) 102, plaintiff sought to recover geese of the wild variety which had strayed from the owner. In granting judgment to the plaintiff, the court pointed out that the geese had been tamed by the plaintiff and therefore were unable to regain their natural liberty.

16 This important distinction was also demonstrated in Manning v. Mitcherson, 69 Ga. 447, 450-451, 52 A.L.R. 1063, where the plaintiff sought the return of a pet canary. In holding for the plaintiff the court stated "To say that if one has a canary bird, mocking bird, parrot, or any other bird so kept, and it should accidentally escape from its cage to the street, or to a neighboring house, that the first person who caught it would be its owner is wholly at variance with all our views of right and justice."

17 [4] The Court finds that Chester was a domesticated animal, subject to training and discipline. Thus the rule of ferae naturae does not prevail and the defendant as true owner is entitled to regain possession.

18 The Court wishes to commend the plaintiff for his acts of kindness and compassion to the parrot during the period that it was lost and was gratified to receive the defendant's assurance that the first parrot available would be offered to the plaintiff for adoption.

Judgment for defendant dismissing the complaint without costs.

Note. Paragraph numbers have been added.

FIGURE 1. Sample Case Used in Identifying Case Elements

research on legal writing largely investigates such issues as the nature of legal writing, the process by which professional legal writing is done, the plain English movement, methodology, and writing in the law firm (Rideout, 1991), which gives little guidance for teaching EALP writing. Moreover, investigation of research papers in other fields usually does not apply because legal research papers have some distinctive characteristics. The lack of existing materials and relevant research has compelled us to do our own discourse analyses, which, in turn, has enabled us to prepare appropriate materials. All of the materials for this course are based on research done in-house on published student legal research papers known as *law review notes* (Feak, Reinhart, & Sinsheimer, 2000). One important finding of this work is that the widely known create-a-research-space (CARS) model (Swales, 1990) for research paper introductions in the sciences could not be imposed on legal research papers. Instead, a modified version of the model for legal research paper introductions, shown in Table 3, seemed more appropriate. As can perhaps be inferred from the table, students need a general understanding of problem-solution texts as well as knowledge of how to attract a readership, introduce a legal problem, and construct a road map that both reflects their purposes and facilitates selective reading (Feak et al., 2000). The move-step analysis presented in the table has led to the development of teaching materials for problem-solution texts, metadiscourse, qualification of claims, reporting verbs, and the language used to introduce a legal problem.

Interactive Speaking and Listening

Processing Legal Materials includes opportunities for students to participate in oral case discussion and analysis; Interactive Speaking and Listening, which meets 5 hours per week, focuses on formal oral presentation. This emphasis allows the instructor to work one-on-one with students on aspects of their spoken language. The curriculum is patterned after our EAP academic speaking class so that an EAP instructor with little legal background can teach it. The main goals of the speaking portion of the course are to increase students' awareness and accurate use of (a) different speech types, (b) linking words and organizational signposts, (c) visuals, and (d) legal vocabulary.

TABLE 3. MOVES IN LAW REVIEW NOTE INTRODUCTIONS

Move	Purpose
Prefacing move	Prepare the reader for the discussion by providing pre-text material such as an epigraph, picture, photograph, or cartoon, which may or may not be vital for the understanding of the note
Move 1	Establish a research territory
	a. by introducing the topic in a general way with some form of opener and showing that the general research area is important or problematic (obligatory)
Move 2	Establish a legal problem or issue (the niche)
	a. by providing some background for the legal problem (obligatory)
	b. by indicating, either explicitly or implicitly, a weakness or problem in the law (obligatory)
Move 3[a]	Address the legal problem and state the nature of the argument (occupy the niche)
	a. by outlining purposes or summarizing the present research (obligatory)
	b. by indicating the structure of the Note through a roadmap or overview (obligatory)

Source: Based on Feak et al. (2000, p. 203).
[a]Moves 3a and 3b may be reversed.

The listening component of this class is like a traditional EAP listening class except that listening materials have a legal focus. Using videotapes from the law school, students gain experience in understanding first-day lectures (including professors' expectations and other practical course matters), regular law lectures and discussions, and in-class briefings. These videos of actual law school classes are valuable because they expose students to the different teaching styles of the professors, demonstrate the demands placed on law students, and reveal the challenges of note taking in classes where discussion, as opposed to lecturing, is commonplace. In Processing Legal Materials, students read the cases discussed in videotaped classes so that they can follow the tapes more easily. By watching the law class videos and observing the sometimes lengthy questioning of students, along with the nature of those questions, the students become familiar with the law school class environment. In addition, students learn to recognize the shift from talk about assigned cases to discussions of hypothetical cases and pending cases. Finally, the students have an opportunity to compare their own discussion of a case with that of the native speakers on the videotape.

Evaluation

Because most students elect to take the ESP program on their own, we do not award grades. Students in general are hardworking and highly motivated, so there is no need to give grades as an incentive. We give considerable feedback, however, so that the students know their strengths and weaknesses. In the writing-focused classes, students meet individually with the instructor to review the written comments on each paper. Students are videotaped when (and sometimes before) they give presentations and again receive individual feedback. This individualized feedback is often cited as one of the most valuable aspects of our program.

Students have an opportunity to evaluate the program and teaching staff at the end of the summer session. Through multiple-choice and open-ended questions, students give both feedback and recommendations for change, which we take into consideration when we begin working on the program for the next summer. Changes that have been implemented as a result of student feedback include reading more cases in business law and changing the research topic of the research class from a comparative legal issue to any legal issue that the students are interested in investigating. Overall, response to the program has been extremely positive and encouraging. In fact, many students recommend the program to colleagues who will be entering an LLM program in the future.

◈ DISTINGUISHING FEATURES

Focus on Authentic Academic Legal English

The ESP program acknowledges the need for presessional instruction for students undertaking degree programs that demand full-time study and allow little time for supplemental language instruction. What sets us apart from other ESP programs, however, is our emphasis on academic legal English. Because of this focus, all the reading materials are authentic, and all the writing tasks are similar to those assigned in a typical LLM program. Thus, students gain experience with the kinds of tasks they will eventually face. Additionally, our small classes allow opportunities for frequent one-on-one instruction, which is not always feasible in programs with large classes.

Legal Expert as Instructor

The fact that two ESP classes are taught by an EAP instructor who is also a lawyer is a critical feature of the program. As an insider to the field, the instructor not only can shed light on the law school community from firsthand experience but also has the content knowledge to discuss the cases and guide student reading. Most of all, the instructor can model the thinking processes that should underlie the reading of cases. Having a lawyer/EAP instructor teaching two key courses, we feel, increases the program's legitimacy in the eyes of the students and their sponsors and has proved to be a major factor in the program's success.

Effective Use of Specially Tailored Materials

Because nearly all of our materials are produced in-house, based on our own research, the materials precisely fit the goals of the program. In addition, because we develop our own materials, we are aware of how we might use these materials in more than one class. For instance, some cases chosen for Processing Legal Materials are also discussed in the videotaped law classes that the students watch in Interactive Speaking and Listening. Similarly, students may read cases in Processing Legal Materials that they will need to write about in Academic Legal Writing. This overlap gives students considerable exposure to and practice with legal materials and the strategies required to deal with them.

Workshop in Languages of the Law

Another distinctive element of the program is the 8-hour, 2-day workshop, Languages of the Law, which takes place about two thirds of the way through the program. The workshop, conducted by two linguists who teach only these 2 days, takes the place of the normal schedule of classes. Whereas the regular courses primarily take a skills approach to the teaching of academic legal English, the workshop aims to sharpen students' awareness of the character of legal discourse by exploring the linguistic features of legal English and examining the ways these features combine in different types of legal discourse, such as briefs, court opinions, and statutes. A main goal of the workshop is to allow students to investigate more deeply the written genres they have been working with in terms of purpose and conventions as well as their linguistic features (Fredrickson, 1998). The workshop includes discussions of syntax and vocabulary, which the students find extremely helpful because they are still trying to sort out the differences between legal and general English.

Although there has been little systematic investigation of legal language (Danet, 1985), the workshop includes description of some of its distinctive characteristics. For instance, students learn about terms of art (e.g., *certiorari*), the use of Latin and French (e.g., the Latin *amicus curiae* and the French *voire dire*), common words with uncommon meanings (e.g., *said* defendant was seen leaving the premises), archaic expressions (e.g., *herewith*), and doublets (e.g., *cease and desist*). Workshop presenters discuss "the degree of appropriateness of choosing these lexical items in writing for a range of audiences" (Fredrickson, 1998, p. 23).

An unanticipated benefit of the workshop is that it gives students a welcome break from the usual routine. Students find time to reflect on their learning up to that point and to spend additional time preparing their research papers. For us, the break provides much-needed time to prepare new materials based on needs revealed during the first 4 weeks of the program.

Law Class Visits

Visits to law school classes are another distinguishing feature of the program. After arranging a visit with a professor and before the visit takes place, we have the students read the cases to be discussed in the class to be visited. By reading the cases ahead of time, the students can follow the class discussion and can see the aspects of the cases that they should have paid attention to while reading. The visits are often a humbling experience for even the best of students because of the fast pace of the class, the nature of the questions (which cannot be answered by just finding the right place in the case), the digressions, and the discussion of hypothetical or additional real cases the students have not read.

Films and Field Trips

Because we are committed to extending students' use of English beyond the classroom, the program includes an extracurricular component. The arranging of extracurricular activities is not unique to our program, but the kinds of activities arranged are. Activities have included a weekly series of feature-length films on law-related topics, such as *Philadelphia, Dead Man Walking,* and *My Cousin Vinnie;* field trips to a state prison; and a half-day visit to U.S. federal court.

After each film screening, the special activities coordinator leads a discussion, which gives students an opportunity to practice their English in a setting that is less formal than the law school classroom. The prison visit is led by a prison official, who takes us to all the main prison areas; students meet with inmates and talk about their situation. Discussions after the visit are interesting and lively as students compare the prison they have just visited with prisons in their own countries. Finally, our courtroom visits have been an exceptional learning experience because the judges are usually as interested in our group as we are in the court proceedings. Our group has regularly been invited into the judges' chambers to discuss aspects of the case being heard or talk about an element of the U.S. legal system, such as juries.

◈ PRACTICAL IDEAS

Obtain Necessary Support

Given the highly specific nature of EALP, establishing such a program poses a number of challenges. The small number of nonnative-English-speaking students studying law (as compared to students entering other fields) means that an EALP program will likely be much smaller than a general EAP program or a business English program and, therefore, may be less profitable. The backing of our institute and the willingness to offer the program even if it was not profitable during its early years has allowed us to nurture it without worrying that the program might be canceled and our efforts wasted.

Select Appropriate Staff

Staffing is also more difficult for our program than for an EAP program because at least the reading course must center on legal cases, which are best handled by a dual professional—an ESP teacher and lawyer. A challenge for most EAP instructors in teaching EALP is that the academic skills that they are generally familiar with, such as presenting clear, concise arguments, participating in class, and reading with a critical eye, take on some unfamiliar characteristics, shaped as they are by the legal discourse community (Feak et al., 2000). Furthermore, because few commercially produced materials are aimed at nonnative-English-speaking lawyers, most program materials must be generated in-house, requiring ongoing research.

Although an EAP instructor with no prior experience in this area may be able to adapt some tried-and-true EAP materials, successful adaptation is dependent on the instructor's having some familiarity with the conventions of academic legal discourse in particular and the legal system in general. Information on the legal system is widely available; however, with the exception of some investigations of case reading (Lundeberg, 1987), little discoursal research has been done in the area of academic legal English. The little research available has traditionally focused on professional settings (Bhatia, 1993; Fredrickson, 1995; O'Barr, 1982). The relationship of this research to the teaching of EALP, however, has been tenuous. As Harris (1992) states, the law school has "largely remained on the margins of EAP work in universities" (p. 19).

Select Appropriate Content

A major challenge for us has been determining whether the content of some of our classes is appropriate. For example, although we have offered a successful presentations skills class with legal content, we realize that we need to undertake further research on legal academic speaking to determine whether the speaking portion of the program should emphasize presentation skills or some other oral skills.

Develop Suitable Materials

Materials development is an ongoing concern for us. Although there is no shortage of legal cases, finding cases that allow us to work on specific reading strategies and to demonstrate ways in which students should be reading is difficult. Developing materials for the writing-focused classes is a challenge because of the time-consuming nature of the discoursal work that needs to be done. In addition, developing materials for the listening component has been somewhat difficult because few law school lectures are taped. Of course, the workshop presenters face the same challenge as the other instructors in that there is a dearth of materials; thus, one area to which they would like to devote more of their energies is the development of more language tasks that give students practice in the language skills they will need in and after law school—a somewhat important consideration for us because many of our students stay in the United States for internships after completing their law degrees. Further materials development, however, is dependent on our continued research of academic legal genres and general legal English corpora.

Modify an Existing Program If a Specialized Program Is Not Feasible

Despite the challenges, designing courses for incoming law students is still possible. If a full-time EALP program is not feasible, one approach to meeting the needs of nonnative-English-speaking law students would be to modify an existing EAP program to include academic legal reading and writing while maintaining the EAP focus of the other courses being offered, such as speaking or listening. In fact, we did precisely that in creating the pilot program before developing the full program. Our pilot program focused primarily on case reading and was very enthusiastically received by the small number of students who participated. Although such a modified program will not expose students to as many of the skills and as much legal language as a full program will, it may still give learners a good opportunity to hone the linguistic skills needed for law school.

◈ CONCLUSION

Although we have encountered a number of unforeseen obstacles in developing our EALP program, centering primarily on our relationship with different parts of the law school, staffing, and materials development, each year we move forward. In addition to full support from our department, we have the backing and cooperation of the law school, in particular the assistant dean of student affairs, as we continue to develop the ESP program.

Although student response immediately after the program has been consistently positive, we have also been especially pleased to learn that after one semester of law school, students feel even more strongly that our program helped them make the transition to the law school environment. Students have reported feeling sufficiently equipped to handle the demands they face. In addition, they continue to offer valuable suggestions for improving the program, such as devoting less time to comparative legal issues because their primary interest is U.S. law.

Each year, we review our curriculum, add new cases, try to find guest lecturers with expertise in the areas of law of interest to our students, and add new materials that emerge from our ongoing research. Nevertheless, we believe the basic framework is sound. Questions remain, however, as to whether we have identified the most important language skills and strategies that nonnative-English-speaking law school students need. Specifically, the lack of informed research in academic legal English continues to pose challenges in curriculum and materials development. Clearly, much work remains to be done.

◈ CONTRIBUTORS

Christine Feak has been a lecturer at the English Language Institute of the University of Michigan, in the United States, since 1988. She serves as the lead lecturer for writing courses for undergraduates, coordinator of summer presessional programs, and lecturer for the writing-focused courses in the ELS program. In addition to coauthoring *Academic Writing for Graduate Students*, she has coauthored articles on summary writing, data commentary, and academic legal writing. Her research interests include the discourse analysis of academic legal genres, the evaluation of nonnative English writing, and the dissertation- and thesis-writing process.

Susan Reinhart has been a lecturer at the English Language Institute of the University of Michigan, in the United States, since 1981 and is the coordinator of the ELS program. She is a graduate of the University of Detroit Law School and is a member of the American Bar Association. Her research interests include the discourse analysis of legal cases, academic legal reading, and academic speaking.

CHAPTER 2

An ESP Program for Students of Nursing

Virginia Hussin

◈ INTRODUCTION

The bachelor of nursing (overseas qualified) program at Flinders University in Australia prepares migrant (i.e., immigrant) nurses of language backgrounds other than English for registration as nurses in Australia and entry into the workforce. As the ESP lecturer, I teach ESP in three modes: as a discrete, credit-bearing language subject; through team teaching with a nursing lecturer; and through supervision of students in clinical settings.

This chapter outlines the context of the program and describes each of the teaching modes. Distinguishing features of the program include the use of genre analysis in task-based learning, a team approach to skills teaching, and my role as the ESP lecturer in clinical placements. This role has presented opportunities for the development of practical strategies and has also enabled the collection of samples of oral interaction that provide insights into aspects of language use that require further research.

◈ CONTEXT

This ESP program was originally developed as a response to an identified labor market shortage of registered nurses in the state of South Australia. It began in 1989 as a 3-month intensive, nonaward bridging program funded by the National Office of Overseas Skills Recognition. In 1991, it was mainstreamed to a diploma of nursing, and then, in the interests of equity, it progressed to a bachelor of nursing in 1993. Students in this program are awarded 2 years' status for degrees or diplomas undertaken in their countries of origin, so they enter at the third and final year of a bachelor's program adapted specifically to their needs. Although all the students are migrants for whom English is a second language, they come from a diverse range of linguistic and cultural backgrounds.

Before entry, applicants require approval from the Nurses Board of South Australia, the professional registering body that assesses the status of nursing qualifications from the countries of origin. The students are then placed in the course through an interview, conducted by one of the nursing lecturers, and a criterion-referenced assessment, which I administer. This assessment is an adaptation for the nursing context of the Australian Second Language Proficiency Rating (Ingram &

Wylie, 1995). I rate the students according to their proficiency in each of the macroskills: listening, speaking, reading, and writing. Ideally, students are at Level 3 (minimal vocational proficiency) at the start of the course and exit as close as possible to Level 4 (vocational proficiency).

❖ DESCRIPTION

The curriculum is structured so that the first semester includes nursing theory classes, taught by the nursing lecturers, and English Language Skills in Nursing 1, which I teach. This class has an English for academic purposes (EAP) focus, set in the context of nursing study. In the second semester, students take a nursing class called Professional Practicum, which includes some theory, an 8-week clinical placement in a hospital, and a team-taught language module. Running concurrently with Professional Practicum is a discrete credit-bearing language class entitled English Language Skills in Nursing 2, which I also teach. This class, the team-taught language module of Professional Practicum, and the supervision of students in the clinical placements all have an English for vocational purposes (EVP) focus, as they provide the students with opportunities to learn and practice the language skills necessary to engage in nursing practice. The EVP teaching in the second semester is the focus of this case study.

Course Design

To design the ESP program, in 1991 I conducted a target situation needs analysis, as suggested by Munby (1978) and Hutchinson and Waters (1987, p. 19), in consultation with various stakeholders (i.e., nursing lecturers, hospital staff, and the Nurses Board of South Australia) by means of interviews, observations of nurses at work, and data collection from documents and nursing texts. This process revealed the main language tasks and skills that formed the basis of a curriculum document (see the Appendix).

In 1993, after the Australian Nursing Council's (ANC) competencies for registered nurses had been accepted nationally, the teaching team, consisting of two nursing lecturers and me, decided to check to see how well the skills taught in the program matched these competencies. In this process, which was similar to the method used by Svendsen and Krebs (1984), we accompanied three nurses for 3 days as they performed their duties in their workplaces, noting the language tasks and underlying skills. We then matched these skills to the ANC competencies to arrive at corresponding language competencies (Blackman, 1993). Next, we checked our perception of the competency match by consulting with two nurse educators on the hospital staff.

More recently, the course has moved toward the concept of participatory needs analysis as outlined by Robinson (1991): "Where students have already started their specialist studies, they can report to the ESP teacher on the needs which emerge in the course of those studies" (p. 15). This approach is particularly useful when students are already in the target situation, as is the case with the second half of the language subject, which runs concurrently with the 8-week clinical placement.

Teaching Modes

The three teaching modes of the ESP program (see Table 1) are the discrete language class, English Language Skills in Nursing 2; the team-taught language module of Professional Practicum; and the supervision of students in their clinical placements. Weeks 1–9 focus on informational use of language, and Weeks 10–14 cover more sophisticated, interactive language use in interpersonal settings involving, for example, the counseling of patients and negotiating in team meetings.

The Language Class

I teach the language course in 3-hour weekly workshops that run over a 14-week semester, the second half of which runs concurrently with the clinical placement. Most of the materials used in the teaching are authentic in that they "have been produced for purposes other than to teach language" (Nunan, 1985, p. 38). These genuine samples of language-in-use, collected from hospitals and clinics, include audiotapes of phone messages and oral reports on patients; videotapes of patient interviews and team meetings; and written documents such as nursing history forms, nursing care plans, case notes, and discharge summaries. Although the weekly workshops deal with target tasks underpinned by practical skills, they also have a genre-specific focus.

TABLE 1. TEACHING MODES AND WEEKLY SCHEDULE

	Mode and Topic	
Week	English Language Skills in Nursing 2 (3-hour weekly workshops)	Team-Taught Language Module of Professional Practicum
1	Taking a nursing history	Administering medications
2	Observing and recording vital signs	
3	Reading and writing nursing care plans	Dressing wounds
4	Giving and receiving nursing handovers	
5	Reading and writing progress notes	Lifting and transferring patients
6	Making and receiving telephone calls	Clinical simulation (2 days)
	Semester break	
7	Writing discharge summaries	*Supervision of Students in Clinical Placements*
8	Writing referral letters	
9	Writing incident reports	
10	Presenting patient education sessions	Clinical debriefing sessions / Observation of student interactions with patients and colleagues
11	Using counseling skills	
12	Using assertion skills	
13	Participating in team meetings	
14	Communication strategies	

Widdowson (1979) has discussed authenticity as a quality that becomes realized only through response and concluded that "it makes no sense simply to expose learners to genuine language use unless they know the conventions which would enable them to realize it as authentic" (p. 166). One way to help students become aware of these conventions is through genre analysis, as suggested by Dudley-Evans (1988, p. 28), who stated that the purpose of genre analysis is to describe texts and events in a way that will help the teacher understand conventions, expectations, and features associated with them and then use this knowledge to develop teaching material. An example of the way genre analysis is used in this course is included in the Distinguishing Features section.

Students' performance in the language course is assessed through three tasks: a nursing documentation package based on a patient scenario, a structured role play of a patient education session, and a clinical language learning log. In this log, students describe, evaluate, and reflect on aspects of their language performance during the clinical placement in four main areas: interviewing a patient, delivering a verbal report, dealing with telephone messages, and using assertion skills.

Team-Taught Language Module

As a regular member of the School of Nursing in the Faculty of Health Sciences, I have become immersed in the culture of this discipline by attending department and school meetings, participating in discipline-based research forums, taking part in staff room discussions and debates, and developing professional relationships with health care professionals. I incorporate the insights gained into this particular discourse community into teaching materials and methods. In addition, the team-teaching approach creates a situation in which I can stake a somewhat equal claim to course content and methodology.

Each year the teaching team, consisting of two nursing lecturers and me, meets to assess and interview students for course selection, plan each semester's program, and decide how best to integrate language into the content area. We also discuss how to develop complementary assignments and strategies to use with students who are having difficulty in nursing subjects or in their clinical placements. An example of a team decision was the determination that some areas of language use in nursing interventions (e.g., wound care, the administration of medication, and the lifting or transferring of patients) would be best approached in an integrated way. Consequently, we now team teach these three areas over three sessions as a language module of the nursing class, Professional Practicum. The team-taught language module continues in a more intensive mode through a 2-day clinical simulation the week before the actual placement. Here, the focus is on the contextualized use of language so that students become aware of discourse features and interactional patterns as well as appropriate grammar in authentic situations.

Supervision of Students in the Clinical Placement

ESP experts such as Widdowson (1979) and Bhatia (1993) have written about the need to evaluate ESP courses through students' performance in the target situation for which they have been trained. Assessment in the language course involves three assignments, one of which is based on the clinical placement. Furthermore, as an acknowledgment of the importance of the student's language performance in the target situation, I am involved in the joint assessment of this performance during the

8-week clinical placement, which involves visiting students in their placements to observe them

- interacting with a patient (e.g., admitting a patient)
- interacting with a staff member (e.g., receiving and clarifying a set of instructions)
- conducting a telephone conversation (e.g., making a call to a community agency)

I visit each student in his or her placement, check the documentation—such as case notes or discharge summaries—for accuracy and appropriateness, and discuss the student's progress with hospital preceptors and with the supervising nursing lecturers.

◈ DISTINGUISHING FEATURES

This program is distinctive in its use of genre analysis in task-based learning, its team approach to teaching a nursing skill, its use of simulation and role play in teaching, and the opportunities afforded to identify students' areas of difficulty.

Genre Analysis in Task-Based Learning

Swales (1990, p.13) stresses the importance of communicative purpose within different communicative settings and the effect of the purpose on the structure of different genres. More recently, Bhatia (1993) has used a *thick description* of language to develop further the approach of using genre analysis for teaching. Bhatia describes the process as follows:

> In order to introduce a thick description of language in use, it is necessary to combine socio-cultural (including ethnographic) and psycholinguistic (including cognitive) aspects of text construction and interpretation with linguistic insight, in order to answer the question, Why are specific discourse-genres written and used by the specialist communities the way they are? (p. 11)

Nursing care plans certainly fit Swales' (1990) description of a genre in that they are "highly structured and conventionalised" (p. 13), and I use genre analysis to teach the target task, Writing a Nursing Care Plan. This approach, represented in Figure 1, is an adaptation of Bhatia's model.

Stage 1 in the approach is called Building Knowledge of the Genre. This involves

1. discussing the place of nursing care plans in the overall nursing process, that is, after the taking of a nursing history and the assessment phase and before interventions are carried out or nursing reports are written

2. discussing the purpose of nursing care plans and their relationship to the audience, that is, the other nursing staff and the way the staff uses the nursing care plans

3. referring students to nursing texts on the construction of care plans

4. collecting samples of nursing care plans from the students' clinical placement venues

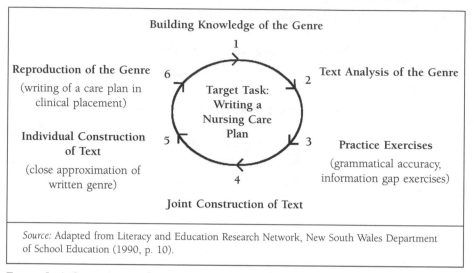

Building Knowledge of the Genre

Reproduction of the Genre

(writing of a care plan in clinical placement)

Text Analysis of the Genre

Target Task:
Writing a
Nursing Care
Plan

Individual Construction of Text

(close approximation of written genre)

Practice Exercises

(grammatical accuracy, information gap exercises)

Joint Construction of Text

Source: Adapted from Literacy and Education Research Network, New South Wales Department of School Education (1990, p. 10).

FIGURE 1. A Genre Approach to Nursing Care Plans

Stage 2, the actual analysis of the genre, involves two steps:

1. looking at the patterns in the genre to see how it is organized on the discourse level. We break down the nursing care plan into its five component parts: the nursing diagnosis, the client-centered objectives, the nursing interventions, the rationale for the interventions, and the evaluation of the interventions. The students determine the intention or purpose of each part and explore the cognitive, semantic, and syntactical relationship among the component parts.

2. focusing on the lexical and grammatical features of each of the component parts. For example, the nursing diagnosis is a statement of a condition describing cause and effect, such as *Ineffective breathing pattern related to decreased lung expansion.* Some of the salient features here are complex and compound noun phrases, connectives, qualifiers, and modifiers.

In Stage 3, Practice Exercises, the students are encouraged to manipulate the text, for example, through information-gap exercises. In one of these, a jigsaw gap, students receive all the elements of the care plan, but out of sequence, and reorder them to form a cohesive unit. To complete the task, students need to be able to recognize and identify the features of a particular component part. In this stage the students also practice writing skills. For example, they might write objectives to match certain nursing diagnoses or write interventions to fit certain objectives. The activities in this stage focus on accuracy.

Stage 4 is a joint construction of the whole text using raw data, done step-by-step in class together. Then, in Stage 5, the students construct their own care plans individually and receive feedback on them. At this point the focus is on fluency. The final stage occurs in the clinical placement, where students write nursing care plans for their practicum and for assessment purposes.

A Team Approach to Teaching a Nursing Skill

Team teaching enhances authenticity in the classroom by involving students in purposeful, structured role plays and extended simulations. We use this approach to teach the nursing skill of lifting or transferring a patient (see Figure 2). In this five-step process, the nursing lecturer first covers the necessary areas of nursing knowledge, for example, the principles of lifting a patient. The nursing lecturer and I then generate a whole-group discussion about the sociocultural context of this nursing task, for example, who can touch whom, where, and when. In some of the nurses' countries of origin, there is little or no cross-gender lifting of patients, a rule that often also applies to other nursing tasks, such as washing a patient. In other countries, nurses do not lift patients at all, as this task is done by orderlies (also called *wardsmen* or *housemen*).

Second, the students twice watch a video showing various lifting techniques. In the first viewing, students observe the lifting techniques, and in the second viewing, they focus specifically on the language used by the nurses and patients in the video. In this way, the video serves as a stimulus for students, as it presents new lexical items and models of language in use. By watching the video, students are also made aware of the stages of the interaction.

The third step involves reviewing the language used in the video from the students' own notes and through a brief analysis of the video transcript. They are encouraged to focus on lexis, speech acts, and structures. The students themselves usually come up with a sequence of speech acts used in the video, such as offering assistance (to lift a patient), requesting help from a colleague (for the lift), explaining the procedure to the patient (especially when using a mechanical lifting device), giving instructions to the patient, checking the readiness of the colleague, offering reassurance to the patient, and checking the patient's comfort levels. We draw students' attention to speech-act shifts, which reflect register variables, such as the shift from giving instructions to the patient to checking the readiness of the

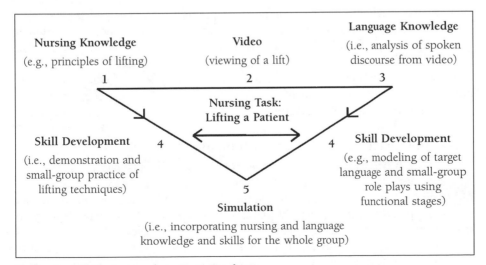

FIGURE 2. A Skills Approach to Team Teaching

colleague. The students need to be able to identify how these shifts are signaled and executed before becoming comfortable with making such shifts themselves.

In Step 4, students develop and practice the physical lifting techniques under the direction and supervision of the nursing lecturer. The students are then encouraged to add language to their actions by replicating some of the target language from the video and by experimenting with some of the student-generated alternatives in small-group role plays.

Finally, the teaching and learning culminate in a longer role play in front of the whole group, in which students must demonstrate a patient transfer or lift while using the appropriate language, functional stages, and speech-act shifts. These role plays often end in a discussion of associated idioms, such as *I've got a crook back* or *I've done my back in*. This session is also followed up in the language workshops by a teaching segment on writing incident reports, for example, in the case of a patient who has accidentally fallen out of bed.

Use of Simulation and Role Play

The course involves a lot of role play and simulation, and we have found very little reluctance on the part of the students to play active and creative roles to acquire communicative competence. This is probably because the students are already in the target situation, so their needs are both "instructional" and "operational" (Strevens, 1988, p. 6).

The 2-day clinical simulation takes place in the week preceding the actual clinical placement. For the simulation, students are organized into nurse-patient dyads and are given different case scenarios to work with. The "nurse"must admit her or his "patient," take a nursing history, prepare the patient for the operating theater, care for the patient postoperatively and prepare the patient for discharge from the "hospital." This pre- and postoperative simulation requires the intense use of all of the language macroskills, for example, listening to verbal reports and nursing handovers, discussing patients' anxieties with them, reading pre- and postoperative orders, and writing at least four different documentation genres. The nurses must also make four telephone calls to different people regarding their patient's welfare.

During the simulation, the nursing lecturer and I take notes on student performance. Each dyad is videotaped when the nursing history is being taken, and at least one of each student's telephone calls is recorded. Most of the teaching occurs in the extensive postsimulation evaluation and feedback session, when we use the notes and recordings to comment on interactions, give feedback, suggest changes and alternatives, and correct grammatical errors.

Opportunities to Identify Areas of Difficulty

My role in supervising students in their clinical placements has presented rich opportunities to identify areas of language difficulty in the target environment. Although students are well motivated, some of them experience problems once they start their practicum in the clinical placement. The kinds of problems depend on many factors, one of which is English language proficiency level. Errors in written documentation, perhaps the easiest to identify, range from word order—particularly with prepositions in phrasal verbs and adverbs of time and frequency—to preposition choice. Verb tense can be problematic, especially the use of the past versus the perfect in case notes that are usually written under pressure just before the *handover*—the verbal report given from the staff of one shift to the staff of the next. In other types of nursing documentation, such as nursing history forms, the verb tense required is more predictable, and there is usually more time to complete the task.

A more significant area of difficulty for many students is communicating over the telephone because they do not have nonverbal clues as to what is being said and often have to deal with background noise, distractions, and interruptions. Early in a placement, students have sometimes been observed walking briskly away from the nursing station when the telephone starts to ring or pretending it is not actually ringing. Many students have particular trouble identifying numbers and names of people over the telephone and distinguishing between intonation patterns for various clause types, for example, declarative versus interrogative (i.e., distinguishing between an interlocutor's statement or question).

Another area of challenge for some students is pronunciation, particularly distinguishing between initial and final consonants (e.g., in the words *Betadine* and *Pethidine*). In face-to-face interactions, students may also experience difficulties with word order in phrasal verbs, preposition choice, and the use of tag questions. In the following transcribed conversation, the student is having difficulty understanding the way verbs can be used in context, but she successfully asks for clarification.

Preceptor: She's having an appendectomy.
Student: She's having it *now*?
Preceptor: No, she's *in* for an appendectomy. She's having it tomorrow.
Student: Oh . . . She will have it tomorrow.
Preceptor: Hmm [nods her head].

Selinker (1988) notes that examples of implicit rhetorical classification, such as generality and conditionality, are more likely at a specialist-to-specialist level. We have certainly observed this to be the case when comparing nurse-to-nurse interactions with nurse-to-patient interactions in the hospital placements. Students must often mentally fill in elliptical statements or questions, as in these examples.

Student: Could you do my aseptic technique assessment next Tuesday?
Preceptor: Well, actually . . . Tuesday's my RDO [rostered day off].
Student: . . . er . . . what about Wednesday?

Preceptor: Now that you've seen Mrs. Mitchell's care plan, what do you think?
Student: [looks blank]

Doctor: [to patient, on rounds with student nurse, looking at the patient] We're going to have to do some more blood tests today.
Doctor: [to student nurse, 10 minutes later] Have you got those path forms ready for Mr. Lee?

In the first example above, the student successfully fills in the ellipsis. In the second, the preceptor's meaning is more obscure. The preceptor could have been referring to the care that Mrs. Mitchell had been given, Mrs. Mitchell's current health status, or even the quality of the care plan itself. In fact, the preceptor was asking the student to give her opinion about what interventions should be implemented and what changes should be made to the care plan.

In the third example, the doctor's original statement, addressed to the patient, is actually a cue for the nurse to prepare the pathology request forms. In this case, the student did not pick up this cue (although she reported that she picked up a similar one from the same doctor 2 weeks later). It is debatable whether the doctor should expect the nurse to pick up his cue or, as a matter of courtesy, make his request to the nurse quite clear. This situation relates to how power is expressed in language and to culturally based expectations and perceptions of the nurse's role.

Idiomatic English can present students with particular challenges. For example, some idioms could have literal meanings in a health context, such as *It's no skin off her nose!* The Australian habit of shortening words can also present difficulties for students, as in this next example. A patient who had recently given birth was admitted for an emergency removal of gall stones and was at this stage postoperative.

Nurse: You'll need to check her peri [perineum].
Student: Her . . . peri?
Nurse: Yeah, she had a pisi [episiotomy] with the birth.
Student: Sorry? I don't understand.

The student knew the words *perineum* and *episiotomy* in their full form, but students will be expected to understand shortened versions as well as the more common, conventional abbreviations used in hospitals. Although colloquial language and abbreviations are taught in class, the instruction cannot deal with every example that students may encounter, so we emphasize the importance of using clarification techniques.

Finally, in face-to-face interactions, students may be challenged by the function of emphatic word stress and tone, particularly when used in sarcasm. For example, when a nurse, referring sarcastically to a visiting specialist, commented to a student, "Don't you just love his bedside manner?" the student smiled nervously and later reported her uncertainty as to the nurse's meaning in class.

◈ PRACTICAL IDEAS

To address some of the initial difficulties that students experience in their placements, we have instituted a range of practical strategies. Some could well be adapted by ESP educators working with students in practicum-based courses in medicine, physiotherapy, pharmacy, social work, and teacher education. The aim of these strategies was to create a supportive learning environment for students in the belief that "too often communication problems are seen as purely the incomer's problem" (T. Boswood, personal communication, 1995), a view that is reflected in recent moves toward a more critical target analysis.

Establish a Preceptor System

The first step taken to create a supportive learning environment in the nursing program in which I teach was to establish a register of committed hospital preceptors—people who have demonstrated that they share the philosophy of the course and value the bilingual and bicultural backgrounds of the students. The establishment of a one-on-one preceptor system exposes the student to a consistent language model and allows for a safe environment for discussion and for the preparation and rehearsal of language tasks (Hussin, 1991).

The next step was to organize preceptor workshops before student placement, facilitated jointly by the nursing lecturer and me. These workshops have focused on the language, learning, and sociocultural needs of the students and include written handouts that preceptors take back to the hospitals with them. These handouts suggest ways that staff can modify their own spoken language to make it more accessible to students, particularly in relation to the use of tag questions, negative structures, colloquialisms, and sarcasm. They also cover strategies that the staff can use to help students understand English in the nursing context, for example, by paraphrasing verbal information that students have not understood or by providing model copies of documentation that make expectations more explicit. Hospital staff are asked to give the students specific feedback on their documentation, outlining what their strengths are and also where and how they need to improve. This feedback is passed on to me so that the university classes can reinforce the points of strength and work on weak areas.

The third strategy was to build in support mechanisms during the hospital placement. These mechanisms include ensuring that students have plenty of passive exposure to a language task in the placement before they have to perform it themselves. For example, students first listen to verbal nursing reports or handovers and observe note-taking, then practice giving their own handover to their preceptor, and finally perform in front of the whole nursing team.

The same principle applies to telephone calls. Preceptors often give students practice telephone exercises that require them to receive a message. As their confidence increases, students practice initiating a call involving more negotiation of meaning and are gradually encouraged and expected to make and receive telephone calls in the hospital. To cope with telephone conversations, students are trained in the use of clarification devices, which are crucial for safe nursing practice.

Institute a Reflective Practice Model in Debriefing Sessions

Perhaps one of the most important aspects of supporting students' learning in the hospital placements is the implementation of a reflective practice model in the weekly debriefing sessions. The students come to the university on the first day of every week and stay in their placements for the remaining 4 days. In these sessions, students are encouraged to suggest learning tasks based on areas of need that they have identified during their placement. In this way, the target situation provides the framework for learning activities. Students can develop language skills in class and then try them out in the clinical setting that week. The following week, students can report back on their success in class debriefing sessions and, if necessary, refine skills before applying them again in the hospital setting.

As the placement progresses, the focus of the extended debriefing sessions becomes the development of strategic competence, using data provided by the students as they relate their experience of language use in the clinical setting. This is done through *playback*, a technique borrowed from counseling, but here the students use it to identify breakdowns in communication. During the debriefing session, students identify a difficulty they have experienced in the previous 4 days of placement and describe the interaction, which I script. The students explore other linguistic choices and strategies that could be used to achieve the desired outcome and incorporate them in a reenactment of the scenario. This technique is particularly useful when students reflect on interactions requiring advocacy, assertion or negotiation.

Students are also encouraged to develop sociolinguistic and strategic competence through their final assignment for the ESP topic, the clinical language learning log, in which they reflect on three genres of interactions: a patient interview, a verbal report, and a telephone call. Students must describe the interaction in detail, rate its success, identify difficulties, identify strategies used to overcome difficulties, evaluate the guidance or feedback they received, and identify ongoing learning needs.

Use Data From the Placement to Determine Future Research Directions

This program has proven to be an effective pathway for nurses from language backgrounds other than English to enter the workforce. In debriefing their clinical experience and collecting data on their own language performance, the students are in part learning skills of ethnographic analysis that will prepare them for their future working lives, or, as Swales (1990) expresses it, "in learning to analyze participant roles and sociorhetorical conventions they are preparing to survive the *rites of passage* of new discourse communities" (p. 218). However, an analysis of both student-collated debriefing data and preceptors' comments on evaluation forms reveals the need for further research into how language is used in interactions in clinical settings.

For example, preceptors frequently comment that students have trouble keeping up with verbal instructions, which is not surprising, as nursing staff often make rapid speech-act shifts within an interaction. A nurse might move from a request, to a complaint, to a joke, to a set of instructions, all within 3 minutes. Therefore, some relevant research questions might be, How do speech act shifts interact in the clinical area? How are these shifts signaled and executed?

Students also have to deal with degrees of implicitness, for example, with causality and conditionality in verbal nursing reports concerning patients or with modality in the hedging that takes place in negotiations about roster changes. Possible research questions in this area might be, How is implicitness expressed in interactions in the clinical area? How can implicitness be unpacked or revealed by nonnative speakers? How can it be dealt with or responded to?

By far the most important emerging research area is the need to identify and explore degrees of assertion within cross-cultural role relationships. Preceptors complain that students do not take initiative, for example, do not speak up for themselves or participate actively in team meetings. At the same time, students express frustration with their uncertainty about how to get what they want for themselves or their patients, which is particularly crucial when they are acting as patient advocates in interactions with doctors and hospital administration. Important research questions in relation to assertion are these: How is assertion expressed in cross-cultural relationships in the clinical area? What is the impact of culture on the expression of assertiveness in these relationships? How can students be better trained in this particular sociocultural competency?

❖ CONCLUSION

The South Australian course for overseas qualified nurses described in this chapter is distinctive in its use of genre analysis in task-based learning and its team approach to skills teaching. The practical strategies for enhancing the development of students' communicative competence within the clinical placement have wider application for ESP educators in practicum-based courses, particularly in the health sciences.

When students are close to, or in, the target situation, the ESP teacher must get inside the texts and tasks of the specialist language and its purposes in order to exploit learning activities around them, a position supported by Swales (1986), who stressed the need for research into academic, professional and occupational "subcultures" that will reveal important information about, for example, belief systems and value judgments: "The ESP practitioner as 'insider' rather than 'outsider' needs to have some appreciation not only of the conceptual structure of the discipline but also of the conventions of conduct imposed upon its members" (p. 10).

For the overseas qualified nurses, these conventions include the unstated assumptions and sociocultural knowledge shared by members of the health profession, and the roles include those embedded in the institutionalized use of language in clinical settings. Clearly, ESP practitioners need more knowledge in the three areas of speech act shifts, implicitness, and assertion in order to extend students' sociolinguistic repertoire and develop their sociolinguistic, pragmatic, and strategic competence. Indeed, these three areas indicate useful directions for future research into language use in clinical settings.

❖ CONTRIBUTOR

Virginia Hussin is a learning adviser at the Flexible Learning Centre at the University of South Australia and has been an ESP lecturer in the Faculty of Health Sciences at Flinders University. She has taught ESP to scientists in China and to teachers, nurses, and doctors in Australia. Her research interests include interaction analysis in the workplace.

❖ APPENDIX: CURRICULUM FOR THE ESP PROGRAM

Main Language Tasks

- taking a nursing history
- reading and writing nursing care plans
- giving and receiving nursing handovers
- reading and writing progress notes
- writing discharge summaries
- using language in nursing care delivery
- writing incident reports
- writing referral letters
- making and receiving phone calls

- presenting patient education sessions
- participating in team meetings

Skills Inventory

1. Informational Use of English

Interactions With Patients and Their Families

- using interviewing techniques
- giving instructions
- requesting cooperation
- offering reassurance
- explaining medical ideas in easy language
- explaining procedures
- seeking permission
- giving feedback
- comprehending colloquial language
- using teaching techniques

Interactions With Colleagues

- comprehending and giving instructions and directions
- comprehending and giving explanations of procedures
- asking for repetition and clarification
- asking for assistance and explanation
- comprehending and presenting verbal information
- answering and placing telephone calls
- accurately conveying telephone messages
- using appropriate medical terminology
- reading and interpreting routine forms, charts, and instructions
- completing routine forms, charts, and instructions
- reading and interpreting medical records and histories
- completing nursing histories
- reading and interpreting notes and summaries
- writing notes and summaries
- reading and interpreting nursing care plans
- constructing nursing care plans
- reading and interpreting letters and reports
- writing letters and reports

2. *Interpersonal Use of English*

- expressing empathy
- interpreting nonverbal cues
- using attending behaviors
- using nonverbal communication
- using reflective listening techniques
- using clarification devices
- using paraphrasing responses
- using summarizing responses
- formulating assertive responses
- expressing personal opinion

CHAPTER 3

An ESP Program for Students of Business

Frances Boyd

◈ INTRODUCTION

Since the early 1990s, the intensive English language program at Columbia University, known as the American Language Program, has offered a 3-week business English course at its campus in New York City, in the United States. Attracting professional, preprofessional, and pre–master of business administration (MBA) students with a range of backgrounds and nationalities, English for Business attempts to meet their needs by combining academic study of business cases and trends with corporate-style training in business writing and presentation skills. There is an emphasis on business communication skills, vocabulary development, and cross-cultural awareness. The program's location in New York City provides opportunities for students to interact frequently with professionals in the business community.

In this chapter, I first define business English and describe the three types of business English learners and the language programs created for them. I then describe the American Language Program's business English course, which seeks to meet the needs of all three types of learners.

◈ CONTEXT

Business English

Business English is the general term for a multifaceted global movement in ESP with roots in both the academic and the commercial worlds. It is variously defined by the language used in business settings, by the learners' needs, and by the form and content of programs that serve the learners. Each complementary definition sheds light on an important aspect of the movement.

Business English can be defined as a subfield that focuses on the development of communicative competence for business settings, also known as target situations or situated contexts in business (Boyd, 1991). Thus, business English is concerned not simply with general communicative competence but with the specific ability to manage the *delicacy of context* (Richards, 1989) that leads to successful business relationships. This ability to adjust one's language to the situation, which is part of overall professional expertise, can be expressed as competencies (Bhatia, 2000). In the case of business English, learners need to be able to select appropriate language and use it strategically to achieve a particular communicative purpose, and even a particular personal style, in business settings or target situations where English is the agreed-on language of interaction. For pedagogical purposes, these competencies can be further broken down into communicative tasks for business. The list might include, for example, those tasks requiring (a) listening and speaking skills (e.g., giving presentations, negotiating, participating in meetings, socializing, taking part in training, telephoning and leaving voice mail), (b) reading and writing skills (e.g., corresponding by e-mail, fax, or letter; reading business and technical materials; writing reports and proposals), and (c) cross-cultural skills (e.g., understanding the nature of participation expected in meetings, i.e., whether to give a point of view or simply listen to a decision; or understanding the expected acknowledgment of status and authority with a senior person, i.e., whether to be collegial or show deference).

In another sense, business English can be defined in terms of the learners' needs, especially as determined by their relationship to the business world, that is, whether learners are in the workplace or preparing to enter it (St. John, 1996). Because they are familiar with the professional setting, learners with business experience have already identified their need to be able to function in English professionally, using all of their knowledge and sensibilities. In contrast, preprofessionals may know very little about the workplace or about the culture of business, so their needs and wants are related to entry into the profession. A third category includes learners who are taking a study leave from work. They need and want academic English as well as business English to reenter and function in the workplace at a higher level.

Finally, business English may also be defined by the way it is practiced in the wide range of proprietary, academic, and company-based programs that serve business English learners (Schleppegrell & Royster, 1990). These programs encompass a large body of material and many approaches. However, there may be a gap between pedagogical assumptions about business English—say, about the relative importance of various genres of writing and the language used in meetings—and actual language use in the workplace. This gap results both from limitations of research and instructor preparation and from enormous changes in technology and globalization that are affecting communication in the business world (Louhiala-Salminen, 1996; Williams, 1988). The variety of programs that teach business English vary by content, approach, format (e.g., length, intensity), and the competence and experience of instructors. Yet much of this rich and varied activity is difficult to discuss because it is described only partially in the literature.

Business English is generally contrasted with English for occupational purposes, or workplace English, which focuses more narrowly on a job category or industry and often includes line as well as managerial personnel. It is also contrasted with English for academic purposes (EAP), which focuses on the language and skills needed for university study. Finally, business English contrasts with general English, which includes language for daily and social life rather than for a specific professional purpose. However, business English, far from being independent of these fields, has areas of overlap with all of them.

The value of ESP, as Strevens (as cited in Johns & Dudley-Evans, 1993) suggests, lies in its efficiency, relevance, cost-effectiveness, and success. These qualities hold true for business English. That is, as compared with programs in general English, programs that assess and focus on the business learners' specific needs are valuable because they save time and money, keep up motivation and interest, and are ultimately rated more useful by learners. Moreover, companies may see business English programs as a good investment if they help prevent costly cultural misunderstandings (Inman, 1985). Also, for participants, another value lies in the potential of these programs to provide opportunities for networking, both among learners and between learners and professionals in the business community (Atlas, 1999).

These benefits accrue when business English programs meet the basic criteria for ESP: "the careful research and design of pedagogical materials and activities for an identifiable group of adult learners within a specific learning context" (Johns & Dudley-Evans, 1993, p. 116). However, these criteria are demanding, and not all programs succeed in offering a business English curriculum that is clearly focused.

Learners of Business English

Learners of business English can be grouped into three categories based on their relationship to the business world: professional, preprofessional, and pre-MBA. The professionals, the largest group, include those who are currently working in business. The preprofessionals, the next largest group, consist mostly of undergraduates who are preparing to enter the business world. Finally, the pre-MBA learners are those on temporary leave from the workplace to pursue graduate studies in business. The learners in each group share certain characteristics and needs, and several kinds of language programs attempt to meet these needs.

Professional

Professional business English learners have business experience and some proficiency in English. These insiders in the profession view language training as part of career development. They are well aware that the purpose of studying language is either to meet the demands of their current job or improve their qualifications for future assignments, perhaps overseas. As they already identify themselves as professionals, they are eager to be able to perform, in English, at their accustomed professional level. The literature reveals little about gender; however, a trainer at one large corporation reports that 20–30% of the employees studying English are women (Morrow, 1995).

In terms of content and approach, these learners need language and cross-cultural communication skills. Moreover, they need and want activities that cast them in professional roles and draw on their expertise. Also, those doing or planning to do business cross-culturally may need and want exposure to the unfamiliar, whether Western business culture or the cultures of new and developing market economies. In terms of format, the professional business English learners generally require intensive programs to accommodate the demands of their jobs. They often prefer short, results-oriented learning modules with highly specific learning objectives.

Three kinds of organizations—proprietary, academic, and corporate—offer programs designed to meet the needs of professional business English learners. The largest providers of English language instruction to professionals are the proprietary programs, followed by the academic programs and then by the company-based training programs. A small segment of the market uses private tutors or self-instruction. In the next few years, distance-learning programs offered by universities, proprietary schools, and publishers will be another alternative for highly motivated professional business English learners who prefer the flexibility and convenience of on-line study.

- *proprietary programs:* These programs are offered by for-profit schools that typically contract with companies to provide English language instruction. They include globally known schools with many branches as well as smaller, local schools. Set up as business-to-business enterprises, the format of proprietary programs meets the needs of professional learners: short, intensive sessions with small classes and flexible scheduling. The schools market services that are easy to recognize, describe, and track. Moreover, these services are often described as *English language training,* which is seen as part of the larger corporate training effort common in large companies. However, such programs may have difficulty with business content and approach. For example, one extensive survey for a large multinational corporation (Schleppegrell & Royster, 1990) found that many of these programs teach general English for a business population rather than defining the specific needs of the business learners and designing targeted materials and activities.

- *academic programs*: Many universities and cultural institutions, such as binational centers, offer business English courses within their academic programs. In content and approach, these courses may fit the needs of professional learners. Taught by trained professionals, the courses generally stress communicative competence for business settings, often including a cross-cultural component. However, the format of academic programs—semester-length calendars and inflexible scheduling—makes it difficult for professionals to attend.

- *company-based programs:* Many multinational corporations have developed company-based, or corporate, programs to meet the needs of their employees. Business English may be one component of a larger language training program that includes general English, English for occupational purposes, and sometimes classes in other languages. To serve a range of employees, the programs often combine in-house with contracted classes. In content and approach, the business English classes are geared closely to current and anticipated job-related needs, such as overseas assignments. Learners are known and treated as professionals. Indeed, the language training itself is a recognized part of their professional activity. Also, the format of company-based programs is ideal for professional business English learners: small, intensive classes designed to fit the work schedule ("A Look at English Education," 1999). Still, such programs have the challenge of meeting high demand and remaining cost-effective. Moreover, even customized programs cannot always accommodate the variety of jobs, the variety within jobs (e.g., those requiring both technical and business English), or the unanticipated requirements of future job assignments (Morrow, 1995).

Preprofessional

Preprofessional business English learners are typically undergraduates with little or no business experience and widely varying levels of English proficiency. Still outsiders to the profession, these learners are mainly involved in degree programs whose primary focus is economics, business, engineering, or a related field. As a result, they may have only a general notion of how language study might contribute to their careers.

The needs of novice professionals are vague, almost by definition, so particular content may be less important than skills and attitudes for approaching professional development. In content and approach, preprofessional business English learners need general business English communication skills, specific job-seeking skills (e.g., interviewing, writing a résumé and cover letter), and exposure to business culture (e.g., professional, local, Western). In format, the preprofessionals need classes that fit into the extensive (academic) calendar of the major part of their degree program.

Most of the preprofessional programs mentioned in the literature are run by academic institutions, both public and private. They offer business communication skills and exposure to business culture. However, depending on their location vis-à-vis the English-speaking world, the programs vary somewhat in emphasis.

- *EFL environments:* In areas where English is taught and used as a foreign language, universities offer business English courses for preprofessional students as part of degree programs. The emphasis is on developing proficiency in general and business English. Some schools also offer courses in English for occupational purposes related to industries of national importance. For example, English for travel and tourism is common at Thai universities (Sinhaneti, 1994). Colleges and universities in countries with noncapitalist or emerging capitalist economies may offer a cross-cultural component that provides exposure to a market economy and its impact on business functioning and communication. The literature describes some cases of official institutional partnering with U.S. universities as a strategy for structuring business curricula according to a U.S. model and establishing English as the primary language taught. Examples include a partnership with the national university in Cambodia (Thong, 1995) and one with a private business school in Poland (Paradiso & Pawlowski, 1993).

- *English as an international language (EIL) environments:* In places where English has the status of an international language, many universities offer business English courses as part of degree programs. The emphasis, however, is not on general business English proficiency but on high-level, specialized business communication skills (e.g., interviewing, presentation, writing) and on opportunities to experience the culture of business. Thus, in EIL environments, programs often take advantage of the students' high level of English knowledge to focus on needs related to entry into the business profession. Universities may employ several strategies to help students achieve these goals. At one university in Hong Kong, for example, students can select specialized course work in such job-seeking skills as interviewing, résumé writing, and cover letter preparation (Boyle, 1995). At a school in Singapore, students hear lectures on business communication, then participate in fieldwork projects that are structured to include hands-on practice in interviewing, report writing, and presentation skills (Yin & Wong, 1990).

- *ESL environments:* In areas where English is the primary language, a small number of universities offer intensive business English programs for preprofessional students, typically 3- to 4-week summer courses arranged for groups. Though they may need general business English proficiency, participants in these programs usually want exposure to business culture. Thus, in ESL environments, the emphasis may shift to exploring the environment, using the classroom as a point of departure. To maximize exposure to the local business community, program designers often organize class work as practical support to fieldwork. In two examples cited in the literature (Murphy & Pascoe, 1996; Yogman & Kaylani, 1996), students from EFL environments learn to function in a business environment in the United Kingdom or United States by performing actual tasks—making appointments, conducting interviews, meeting deadlines, collaborating—required by a field research project. As the fieldwork progresses, class time is used to reflect on cultural differences and to build some competence in business English.

Pre-MBA

Pre-MBA learners of business English usually have business experience and advanced proficiency in English. They may be planning to apply to graduate school, may be admitted to graduate school, or may already be enrolled in a graduate program during the academic year. Some have taken time off from work to attend; others have quit a job to retool for a better position. Most of these students attend intensive summer courses at universities in English-speaking countries, typically on U.S. campuses.

Regarding content and approach, the needs of pre-MBA business English learners are academic as well as work related. In graduate programs of business administration, students need such EAP skills as the ability to participate effectively in case analysis and seminar discussions (Basturkmen, 1999; Micheau & Billmyer, 1987), write highly structured expository prose (Canseco & Byrd, 1989), listen and take notes, read extensively and critically, and do research.

For reentry into the business world, these learners also need and want job-seeking, presentation, and negotiation skills. Perhaps they need to appreciate how the academic skills may overlap the business communication skills. For example, the nuances of turn taking or politeness in a seminar might well transfer to a meeting or a negotiation. Also, pre-MBA business English students need exposure to business cultures other than their own. In terms of format, these learners usually prefer intensive courses of several weeks in duration.

Programs designed to meet the needs of pre-MBA learners of business English are generally intensive (4- to 8-week) summer sessions located on U.S. university campuses, typically taught by ESL faculty or a combination of ESL and business school faculty (Iacobelli, 1993; Tomizawa, 1991). The courses vary somewhat in the emphasis placed on communicative competence for academic and business settings. However, they all offer development of language, academic, and business communication skills as well as exposure to cross-cultural issues in business. The cross-cultural experience can emerge through work with students of other nationalities in the class, interaction with U.S. business professionals, or course content. For example, to understand how government regulates business in the United States, students need to know about Equal Employment Opportunity legislation and the issues of affirmative action, sexual harassment, and disability.

◈ DESCRIPTION

In the varied landscape of business English, the American Language Program of Columbia University has developed a course that occupies a niche between the academic and business worlds. In format, English for Business is a summer intensive course on a U.S. campus, yet it is not limited to academic or professional learners.

In fact, rather than specialize, we have chosen to keep enrollment open and to accommodate professional, preprofessional, and pre-MBA learners. In curriculum and approach, the program seeks to meet the needs that all three groups have in common. This choice is a response to student interest and an extension of the open enrollment policy for our other English courses. It has also come about because the Business School has not expressed a need for a specialized program.

In physical location, the program also cuts across the academic and business worlds: It is situated on a campus that sits squarely in New York City. Wall Street, the symbolic hub of a global business and financial center, is a 40-minute subway ride from the university campus. Taking advantage of its location, the course provides opportunities for students to interact frequently with professionals in the business community. In addition, participants benefit from being part of a group that reflects the diversity of today's business world.

Student Profile

Participants in the business English intensive program vary in purpose, national origin, English proficiency, and business background. Some take the course as preparation for postgraduate study; others enroll in order to boost their performance on the job; a few are simply curious about business. About half of the students plan to go on to degree programs at U.S. universities. Of these, several have been accepted at MBA programs at a number of U.S. graduate schools, others are planning to apply, and a few others plan to seek degrees in technical fields. The remaining half of the students plan to return to work. Of these, about half work in management positions in banks and related financial industries; others work in media, manufacturing, fashion, and other fields. Some have public sector jobs or are self-employed. A few students are sponsored by their companies, but most participants pay their own tuition and fees.

Over the years, about 45% of the students have come from Europe, about 35% from Asia, and about 15% from Latin America. Beetham (1999) highlights the context in which international businesspeople increasingly find themselves and the tremendous challenge it poses:

> A German bank may employ a Chinese economist [who was] educated in the U.S., but [who is now] working according to a more hierarchical business culture that he or she may never have encountered before . . . . For the individual, the ability to develop an understanding of how these models differ, and how to succeed within them, can be the key to personal success. (p. 66)

Thus, the fact that the student population is diverse, mirroring the global nature of contemporary business, offers a significant intercultural learning opportunity, one that students come to value as the course proceeds.

Finally, the students vary in English ability and business background. The minimum requirement is advanced proficiency in English, usually measured by a minimum of 550 (paper-based) or 213 (computer-based) on the Test of English as a Foreign Language (TOEFL), a telephone interview, or both. The students' TOEFL scores often exceed 600, yet the scores sometimes do not correlate well with communicative skills. The speaking and listening skills are affected by time spent in ESL or EIL environments, age, and other factors. Many participants have 3–5 years of work experience, often at an executive level; some are midcareer managers with 10 or more years' experience; and a few have very little direct experience of the business world.

Purpose and Design

The business English course has several purposes: to help students practice the language of business (idioms, functions, concepts and industry-specific vocabulary); to develop specific business communication skills (writing, presentation, and negotiation); to give students a chance to experience and reflect on key U.S. and international business trends, practices, and attitudes; and to familiarize students with New York as a dynamic, global business environment.

The course, taught by a senior member of the faculty, has from 18–20 students and meets for an average of 4 hours daily for a 3-week period in the summer term. At this time, university life is in full swing: Many students and faculty members are on campus, and services are fully functioning. Most classes are held in a seminar room at Columbia University's Graduate School of Business. Students are introduced to and use the business school library as well as the cafeteria, lounges, bookstore, and other facilities. They meet business school faculty and students formally, when these individuals come to class as speakers, and informally, when their paths cross in the public spaces of the business school.

Curriculum and Materials

The business English curriculum is designed to meet the needs of professional, preprofessional, and pre-MBA students and to capitalize on the setting. The main materials are organized by theme and by skill. Within the 3-week period, approximately 65% of the time is allotted to thematic units, 25% to business skills, and 10% to other activities.

The thematic units, including cases and trends in business, are multimedia units designed or adapted for advanced ESP students. The skill strands—business writing and business presentations—are a series of lessons taught in a corporate-training style used for development of executives on the job. These strands have a different purpose and pedagogy. Their aim is not only to develop communication skills for the business setting but also to allow students to experience the culture of U.S. corporate training. Unlike the more open-ended, critical perspective developed in the thematic units, in the skill strands the instructor adopts a fast-paced, corporate-style training model with more narrowly defined performance objectives.

Thematic Units: Business Cases

Most of the thematic units take the form of language-centered business cases, narratives of authentic executive decision-making situations that have been specially arranged to maximize opportunities for learners of English to practice language and learn about culture (Boyd, 1991). The instructor writes most of the cases used in the course, but a few are existing cases that have been supplemented with language exercises. Lasting from 4–6 hours, the cases get students to identify, analyze, and suggest solutions to classic business problems faced by real executives in real firms. Some of the problems (and firms) covered in the cases include the way a commodity becomes a brand and is exported (e.g., by Starbucks Coffee Company); the effects of cultural and ethnic differences in the workplace (e.g., at the International Bank of Malaysia, Euro Disney); ways business can be both socially responsible and profitable (e.g., Ben & Jerry's Homemade, Inc.); the politics of international competition (e.g., between Boeing and Airbus); U.S. government regulation of the private sector (e.g., the case of Mitsubishi's Illinois plant and the Equal Employment Opportunity Commission); and the way predatory pricing stifles competition (e.g., the case of Southwest Airlines).

The language-centered business cases are designed to integrate practice in all language skills, with an emphasis on listening, speaking, and vocabulary. As in cases used in MBA programs at U.S. universities, students gather information about a particular problem within a firm, generate options for action, and, finally, take a stand. Unlike MBA cases, however, these language-centered ESL cases are organized for maximum language practice. Students gather data by listening, doing information-gap reading, surveying classmates, and doing Internet research. They interpret and evaluate the data by gathering opinions from various individuals and drawing on the knowledge and experience of the multicultural group. They focus on information about the company, vocabulary, and negotiation. For negotiation skills, students use a framework derived from Fisher and Ury's (1992) work, with cross-cultural notions derived principally from work by Graham (1985) and Neu (1986). Then, in a simulated business meeting, they apply all of their knowledge and language in order to generate solutions to the problem. Many of the cases in the course are supplemented by related fieldwork assignments, field trips, or presentations by guest speakers.

For example, in the instructor-written Starbucks case, students learn how this retail coffee company has introduced European-style beverages and café culture into mainstream U.S. life. By creating a relaxing, pleasant place to socialize outside the workplace and the home, Starbucks has had a significant impact on the daily life of millions and has watched its stock soar. Currently, the company aggressively seeks to export its formula for success, even to Italy, whose café culture inspired the whole enterprise.

To enter into the Starbucks case, students gather data about the company's successful brand-building strategy by listening to interviews with Chief Executive Officer Howard Schultz and by reading company brochures, news reports, Web sites, and other public sources of financial data. These are supported with comprehension activities, in which students go beyond simply gathering information to discuss questions such as

- What does Schultz mean by "romancing the customer"?

- Agree or disagree: Schultz would say that coffee brewing is more of an art than a science.

- Some say that Starbucks wants to be "the McDonald's of coffee." What do you think this phrase means? To what extent does this describe the company's goal?

Vocabulary items are identified and practiced, woven throughout the case, and reviewed in an oral mastery exercise toward the end of the unit. Expressions include idioms (e.g., *go to great lengths, be on a roll, pay top dollar*); functions (e.g., for conceding a point in negotiation: *Granted . . . , You have a good point, but . . .*); and industry-specific terms (e.g., *arabica coffee beans, aroma, barista*).

In a fieldwork assignment, teams of students go into Starbucks retail stores in a variety of New York neighborhoods to observe how the brand-building strategy is applied locally, experience the product, and interview the manager. In class, the teams pool their findings to discuss whether Starbucks has managed to avoid the cookie-cutter approach and make the stores responsive to the needs of distinctive neighborhoods, as its executives claim.

Against this background, students wrestle with one of the company's main issues: applying the brand-building strategy internationally. In a simulated meeting, students represent different executives, negotiating for their own interests within the larger firm. This is followed by a writing assignment: "With another student, write a short report to a top Starbucks executive. Use the price, product, production, and placement framework to write a persuasive summary of your argument for a new location idea." The case ends with a critique of Starbucks' strategy: the potential gains and losses from the point of view of customers, businesses, and cultural influence.

Other types of cases are taught by an invited professor from Columbia Business School. In one case studied in this manner, *International Bank of Malaysia* (DiStefano, 1974), students are asked to identify sources of tension and strategies for managing them in an ineffectively organized workplace that includes executives from diverse ethnic and cultural groups. Before the professor's visit, students focus on understanding the reading and the vocabulary (e.g., concepts such as *quota, expatriate benefits, loan records for credit, foreign exchange data*). Then, the invited professor conducts an MBA seminar class, which includes a question-and-answer period, team problem solving, group presentations, and critique. In a follow-up class with the business English instructor, students identify and react to the professor's style, discussing its implications for pre-MBA students. As a wrap-up, they summarize the analysis and prepare a written response, choosing to practice either letter writing or case analysis. For case analysis, students are introduced to a model based on Canseco and Byrd (1989) and a Columbia internal document (*Analyzing Cases,* 1992).

Thematic Units: Business Trends

Rather than focus on cases about particular executive decisions, some thematic units are devoted to a current trend or idea in business. The ideas in these units tend to cut across many companies and countries. Business trend units are often built around a presentation by a guest speaker with a highly developed critical perspective. Units of this kind have included "European Monetary Union (EMU): Winners and Losers," "The IMF and the World Bank: A Critical Perspective," and "e-Commerce: A New Paradigm for Business."

The unit on the EMU gives students an overview of this historic development in Europe. Students review the history of the EMU, discuss its real and imagined effects on business, and make a prognosis for the future. The unit begins in class. Students pool their knowledge ("Which countries are involved in the EMU? What are the risks and benefits for them?"), then study a video clip dealing with the earlier efforts at monetary union as well as a reading detailing the benefits of a single currency. Besides essential background, these texts include rich conceptual vocabulary for study (e.g., *optimal currency area, parallel imports, transparent pricing*).

Next comes the vital link to the local business community. In a conference room overlooking Wall Street, a Chase Bank executive engages the students in a probing discussion of the impact of the euro, leading them through a detailed analysis of the winners and losers. In a follow-up class, students review and react to the content of the euro discussion, analyze the executive's presentation skills (some are surprised to discover that he is a nonnative English speaker), and practice the idiomatic vocabulary he used (e.g., *be on the same page, a zillion*).

The unit is sometimes expanded with a student-conducted walking tour of major sites in the financial district based on architectural, historical, and business information supplied earlier by the instructor. As they walk around the narrow streets of old New York, pointing out the U.S. Customs House, the New York Stock Exchange, the Federal Reserve Bank of New York, and other sites, students have a concrete experience of references they have read about in class. They can imagine, for example, the wall that actually bordered Wall Street and can see today's young financiers head for the fresh air and harbor view of South Street Seaport to begin the weekend.

Business Skill Strand: Writing

The business-writing skill strand is a three-part series of lessons woven into the 3-week course at intervals. This series has a highly structured, job-related goal: to help students understand and be able to produce persuasive (and accurate) e-mail, letters, short reports, and résumés. The key notion is that U.S. business writing is reader centered; thus, many of the writer's decisions about organization, format, and style are governed by the audience (Davidson, 1994; Poor, 1992).

This culturally specific perspective is developed through examples and exercises delivered by overhead projector and then practiced either on the computer or by hand. Such stylistic considerations as informality, bias-free language, and directness are raised in critiques of students' work. In the business-writing strand, students also have a chance to compare the reader-centered notion and its implications with their own ideas about business writing, a cross-cultural perspective explored by Bhatia (1993), Connor (1996), and Jenkins and Hinds (1987).

Business Skill Strand: Presentation

Like the business-writing strand, the business presentation strand is a three-part series of lessons woven into the 3-week intensive, providing a change of pace. As in the writing skill strand, the purpose is specific and job related: to develop awareness of and confidence and skill in presenting ideas persuasively (and comprehensibly) to critical audiences. Munter (1999), among others, presents concepts and examples that are useful for reference.

In this part of the course, the instructor uses pedagogical practices more often seen in corporate training than in academic settings: models of effective presentations, lists of tips, and lists of useful language. Students perform and then receive feedback from a peer and from the instructor. Peer feedback focuses on overall comprehensibility and a checklist of presentation skills. Instructor feedback covers these categories but goes beyond them to include organization skills and individualized written comments on fluency, accuracy, and pronunciation.

Typical presentation assignments include summarizing and critiquing a current business news story, identifying and solving a business case problem, persuading investors to take a chance on a new business idea, and presenting a business plan. Although some students find seeing themselves on videotape unnerving, the support of a peer and a set of review criteria help students turn these sessions into valuable learning experiences. Feedback on a video clip allows the instructor to point out strengths and weaknesses. As a result, students often rate the presentation skill component highly in the final evaluation, appreciating the personal awareness and communication skills they have gained.

Other Activities

Besides intensive work in thematic units and business skills, students may also participate in an array of other activities: a librarian-guided tour of the resources of the Columbia Business School library, a student-guided tour of the Columbia campus, a day-long simulation of economic and political conflict in the developing world, a panel discussion on how to survive in business school (featuring current students in Columbia's MBA program), and an off-campus social event, such as a picnic in Central Park.

In addition, for those who have inexhaustible energy, there are opportunities for self-guided learning in the language laboratory. The lab has a collection of business-related books on audiotape and feature films, both accompanied by in-house materials focusing on language and culture. For students who want pronunciation work, the lab offers materials for individualized practice.

◈ DISTINGUISHING FEATURES

Because the American Language Program's business English course is just 3 weeks in length and because it serves such a variety of students, we have had to develop a curriculum that fits the needs that these learners have in common. We have tried to look at diversity as a strength, creating materials and tailoring activities that maximize learning within the group.

A Language-Centered Case Method

A case method tailored to business English students is the dominant pedagogy in the business English program. This language-centered version of the case method gives students the opportunity to practice executive decision making, negotiation, and problem solving, communication skills that are at the heart of both graduate education in business and executive functioning in the business world. These cases are original materials offering up-to-date and relevant settings and problems. While providing practice in all four basic language skills and vocabulary, the language-centered case method offers authentic cultural background and a chance to reflect on cross-cultural similarities and differences.

Links to the Local Business Community

The course encourages students to interact with professionals in the business community through contacts on campus and in the workplace. The field visits are varied and highly structured. Students observe, interview, conduct tours, listen to presentations, and participate in discussions. Each event is preceded by and followed up with activities focusing on content, language, and business communication skills.

Academic and Business-Oriented Approaches to Learning

Attracting professional, preprofessional, and pre-MBA learners, the course combines activities and perspectives characteristic of both the academic and the corporate-training environments. Students are exposed to the case-style teaching and discussion so prevalent in graduate business schools. They also experience the performance-centered sessions in business writing and presentation skills that characterize corporate training. Both approaches build knowledge, explore attitudes, and help participants express themselves more fully in English.

Explicit and Implicit Teaching of Western and U.S. Business Culture

A critical approach to cultural issues informs nearly every aspect of the program. The instructor tries to choose case problems and speakers that deal with such key aspects of U.S. business culture as equal opportunity, philanthropy, and entrepreneurship. In business writing, the cultural implications of bias-free language, informality, and a reader-centered strategy are brought up for discussion. Even the style of teaching and learning—one that often requires students to participate actively and requires teachers to guide and facilitate rather than lecture—incorporates value-laden behavior that has implications for students' understanding of Western and North American business culture.

Critical Thinking and Collaboration Viewed as Professional Skills

The program reflects the view that critical thinking is essential for both language learning and professional development. As a result, activities that stimulate curiosity, originality, and engagement are integrated throughout the business English course. Before studying a case, for example, students predict and hypothesize. Later, they evaluate their original notions and justify any changes in their thinking. When studying vocabulary, students are encouraged to go beyond the dictionary definition to consider connotation, domain, and register. After listening to a speaker, students might analyze the speaker's presentation skills and comment on how particular language and behavior express attitude and intent.

In collaborative activities, students learn to function as an intercultural group while building professional interpersonal skills and cross-cultural awareness. Team and project work mimic the organization of work in many companies. In the program, students participate in group projects requiring discussion and reporting, do paired language practice requiring coaching and feedback, and plan and participate in social events.

Continuous Individual Assessment Through Coaching

In a 3-week intensive course, instructors can assess individual needs in language production; they cannot promise to meet them all. With written feedback on speaking (i.e., fluency, accuracy, and pronunciation), for example, students can discover their strengths and weaknesses. With oral, written, and videotaped feedback on presentation skills, students inevitably discover areas where they would like to improve. In informal conferences, instructors can sensitize students as to how certain behavior, say, too little or too much assertiveness, may be viewed in a business context. Throughout the course, students receive balanced feedback (i.e., on their strengths and weaknesses) on their performance by means of a rubric that coaches through description instead of dictating through prescription. The insights gained from this process can guide students' efforts far beyond the limits of the short course itself.

A Diverse Student Mix

The mix of students, who differ in purpose, nationality, English level, and business background, is used constructively throughout the course. The cultural and linguistic diversity is viewed as a reflection of the global business world. The range of business and academic experience means that professionals can provide experiential knowledge from the field while pre-MBA students may offer insights into business from a more theoretical perspective. In sum, the variety of students offers an opportunity for students to negotiate the linguistic, cultural, and professional landscape they may encounter at work.

◈ PRACTICAL IDEAS

The following suggestions for designers of business English courses emerge from the program's more than 12 years of experience working with business English learners.

Continually Assess Learners' Needs and the Learning Environment

Programmatic decisions flow out of the needs of the student population as well as out of the instructors' sense of best professional practice. In the early stages of design, program developers might consider what students think they need, what outside sponsors may expect them to get, and what is possible with the time and resources allotted. Once in session, instructors should create opportunities to give balanced feedback to the learners and receive feedback from them. At the end, course evaluations are useful for a summative view of the process.

Assessing the learning environment is particularly important in EFL and EIL contexts, where students and instructors may not readily recognize opportunities for exposure to the local business culture. The instructor can structure interactions with the business faculty and business community—such as observations, interviews, interactive presentations, and walking tours—some of which may take place in the first language (in EFL or EIL contexts) but are then reported, analyzed, and written up in English. Networking and doing research are essential for a contemporary businessperson; thus, field assignments provide much more than practice with the language.

Develop an Interest in the Business World

The instructor, who presumably has more expertise in teaching language than in doing or teaching business, should expect to enjoy reading and listening to business news and commentary. This practice is a source of curriculum ideas, an activity to have in common with students, and a good learning strategy (for language and business).

Explore a Language-Centered Case Method

The language-centered case method is suitable for many levels of business English learners, a wide range of business decisions, and many types of companies. Instructors may want to adapt existing cases to meet the needs of ESL, EIL, or EFL students. Cases are available in various forms from many sources: long, dense texts from Harvard Business School Publications (see http://www.hbsp.harvard.edu/); shorter, less complex texts from such textbooks as those by Barton (1993) or Perreault and McCarthy (1999); or cases created for business English learners, such as those by Boyd (1994). Or instructors may choose to research and write their own cases, particularly about locally known firms or international firms with local subsidiaries or locally available products. Many companies, when contacted directly through their public relations departments, will gladly send press kits and other materials free of charge. Also, the Internet is a rich source of timely articles as well as audio material.

Design Thematic Units Based on Business Trends

Instructors should follow business news in order to spot trends and find effective guest speakers who can enrich the class with their experience. In addition, instructors should try to combine audio and video materials with newspaper and magazine articles to create informative and attractive multimedia packages. Every attempt should be made to exploit the material in an integrated manner, for practice in language skills, vocabulary, grammar, and culture. This integration contributes to language learning because it presents language in the most authentic and motivating manner, with the greatest number of opportunities for natural redundancy.

Investigate the World of Corporate Training

By exploring professional organizations, workshops, publications, and people in the field, language instructors can get a sense of how the world of education intersects with the world of business and how business educates the members of its own community. This perspective adds authenticity and interest to more academic formulations. Instructors may want to begin by contacting the TESOL ESP Interest Section, which has established a language training forum within the American Society of Training and Development, a major professional organization for corporate trainers.

Alternatively, instructors who know the world of corporate training well should investigate academic preparation for business. Business English instructors may want to become familiar with curriculum through course descriptions, required texts, and conversations with students and professors. Or they may choose to audit or take a course to observe the pedagogical approaches and familiarize themselves with the content. Finally, the library is an essential resource for perusing materials, meeting students, and becoming familiar with on-line research tools.

◈ CONCLUSION

Business English, like business itself, is growing and changing, and this case gives but a glimpse of ESP in this field. Clearly, however, business English learners and employers value the concept of instruction focused on the language of targeted situations and the needs of specific students.

The American Language Program's business English course is one response to the needs of a varied group of learners. It attempts to identify common needs, exploits the diversity of the group, and draws on local resources. Judging from this case, Strevens' (as cited in Johns & Dudley-Evans, 1993) insistence on the importance of needs assessment and discourse analysis as key design considerations for programs in ESP is valid. Clearly, what we do in the classroom is important. However, we have also realized that, for these learners in this situation, what we do to facilitate students' connection to each other, to the Business School, and to business professionals in New York City may be even more important.

To improve ESP in business contexts further, the field needs more research on the language of targeted business settings, on learners' needs and wants, and on how materials can incorporate new understandings. In the literature, the most consistent call is for more research into the area of cross-cultural communication. As always in business English, insights will no doubt come from both the academic and the business worlds: from researchers and instructors in the university who are investigating ESP, cross-cultural communication, and the impact of technology on communication; from colleagues in the training community who work directly with learners in the field; and from designers and writers of materials in the new media. From the practitioner's point of view, the most fruitful ideas may emerge from the paths that cut across these interests.

◈ CONTRIBUTOR

Frances Boyd teaches academic and business English as well as graduate courses in the TESOL Certificate Program in the American Language Program at Columbia University, New York, in the United States. She designed Columbia Interactive's series of on-line courses in American business writing for nonnative speakers. She is the author of *Making Business Decisions* and *Stories From Lake Wobegon* as well as coeditor of the five-level academic English series *NorthStar* (all from Pearson Education). Her articles on ESP for business and on professional preparation of international teaching assistants have appeared in *TESOL Quarterly* and *English for Specific Purposes*.

CHAPTER 4

An ESP Program for Students of Tourism

Simon Magennis

◈ INTRODUCTION

Although many people are motivated to learn a foreign language because they believe it will help them with their careers in the airline, hotel, or travel industry, there is surprisingly little literature on the subject. In addition, textbooks for tourism English, as opposed to titles aimed exclusively at, say, hotel staff, appear to constitute a relatively new area of the publishing market. Virtually every general English textbook recognizes the importance of tourism, with significant sections devoted to travel, hotels, and restaurants. However, the professional literature has not adequately defined or addressed the topic of ESP in the tourism industry, nor have many English language programs and materials been created specifically for this area.

This chapter attempts to address these needs by describing English for tourism professionals (ETP) in greater detail and documenting 3 years of teaching English to students of tourism at degree level in Portugal. The program gradually moved from dependence on inadequate published textbooks to developing and using "home-grown" material that was more closely attuned to the specific needs of this particular group of tourism students. Although the program is still a work in progress, many of its elements are well established and may be applicable in other contexts.

◈ CONTEXT

Global

The tourism industry is growing in two ways. First, as more people worldwide have an ever-increasing amount of leisure time along with growing disposable incomes, many choose to spend some of this time and money on tourism. The second kind of growth is a combination of semantics and statistics. Activities that were previously seen as unrelated are being recategorized under the umbrella of tourism. Economists are using more sophisticated tools to separate tourism revenues from other revenues, and governments have thus begun to pay greater attention to tourism once they realize how much it contributes to the economy.

The major difficulties in defining ETP can be traced in part to the difficulty of defining *tourism*. When is a restaurant or bar simply a local facility, and when is it a tourism business? Are air traffic controllers part of the tourism industry? What about paramedics at the first-aid post at a popular beach on the Mediterranean? Whatever the status of paramedics or air traffic controllers within the tourism industry, the language requirements are too specific to have any place in a course that is not exclusively aimed at that occupational group. As for restaurant staff, their language requirements are not really particular to their profession; after all, the waiting staff and clients are communicating about the very basic topic of food.

The same issue surfaces in many areas of ETP—hospitality, transport, amenities, facilities, and attractions. Every area is essentially ordinary, and the associated language skills are at first glance equally so. However, ordinary language skills are not necessarily easily acquired, as is illustrated by the fact the almost every business in the world insists on training its front-line staff in, for example, how to answer the telephone or greet a customer. The transactional language associated with any service is ordinary language, well chosen and extraordinarily well executed. "May I help you?" said with a smile, and "What do you want?" both indicate that someone intends to serve, yet the first option is clearly preferable. In the Description section, I propose a tentative definition of ETP in an effort to bring more clarity to these issues.

Local

The School

The Instituto Superior de Assistentes e Interpretes (ISAI) is a private, fee-based college in Porto, Portugal, a city of about 1.5 million people. ISAI awards 3- and 4-year degrees (BA and BA honors equivalents) that are recognized by the Portuguese Ministry of Education and serve as professional qualifications for those seeking a career in the tourism industry. To become accredited, the ISAI and other third-level institutions submit to the Ministry of Education a formal statement of the program content drawn up by the academic council. The statement is broad enough to allow language teachers and academic coordinators to interpret it in light of the actual classroom situation; it does not prescribe detailed content for our language classes and thus allows staff to develop new material independently. In this context, levels or standards, as well as actual content, are ultimately the responsibility of the staff directly at the chalkboard.

Students

In Portugal, mastery of English is fairly easy for those who are interested in the language. As in most Western European countries, English is the predominant foreign language in the school system and is frequently heard on television and at the cinema. Portuguese students have fewer difficulties with English than their peers in Spain because the broad Portuguese phonetic system enables students to learn an accentless English with few pronunciation difficulties. In addition to these learning advantages, a significant minority of Portuguese students have undergone primary and secondary education partially or entirely in other countries, most commonly France or Canada. Consequently, some students in Portugal know English quite well, and a number of them enter our program.

Most students who are studying at the tertiary level in Portugal have studied English for at least 6 years in secondary school. Many have also attended additional classes in the numerous private language schools that are widespread throughout the country. Despite this, language levels among students at the ISAI have varied considerably from year to year. As all classes, including language classes, are organized by year (or semester) rather than by level, this variability presents a challenge for language teachers.

Entrance to tertiary courses in Portugal requires that a student pass the 12th grade (*titularidade do 12° ano de escolaridade*), which in theory allows entry to all courses at any university or third-level college. In practice, places in popular courses, especially in the low-cost but high-status public universities, are limited and awarded to students according to their examination scores. Private, fee-paying colleges are often viewed as second choices to public universities, as entrance is not usually limited once students have satisfied the minimum requirement of a 12th-grade education.

Tourism studies in Porto pose a curious case. As public universities in the area offer no tourism courses, private institutions are the only choice available to students wanting to study for a professional degree in tourism. The result is a diverse student body in terms of language competence. Some students with high scores and correspondingly good language skills are strongly motivated to develop careers in tourism and select the program as their first choice, but others enroll in the program because it is one of the few higher education options open to them due to their low examination scores.

ETP

Although various well-known and widely accepted de facto standards exist for general English—such as the Test of English as a Foreign Language (TOEFL) and the University of Cambridge Local Examinations Syndicate (UCLES) tests—there are few well-known or widely accepted standards for professional English. Where tests exist, how they relate to other tests in terms of general language level is not always clear. To my knowledge, only the London Chamber of Commerce and Industry (1995) offers tourism-specific examinations: the two levels of the English for the Tourism Industry examination. The higher level examination is meant to demonstrate that the candidate has achieved the competencies necessary to work in a variety of tourism positions. However, at Level 5 on the English Speaking Union's 9-level scale (West & Walsh, 1993), it falls short of the English level required for the positions that higher education graduates frequently aspire to. The only other tourism-specific English language examination to which I have found any reference (in West & Walsh, 1993), the Oxford Tourism Proficiency Examination, was discontinued a number of years ago. Some countries that have a vocational stream within their secondary education system include foreign languages for tourism, so some local definitions may exist.

The World Tourism Organization is involved in the education of tourism professionals; however, it has apparently not been very active in the area of language curricula and seems to be more concerned with the management and marketing elements of tourism curricula. Although in Portugal 21 third-level institutions, ranging from public universities to third-level colleges to hospitality schools, offer tourism or hospitality-related degree courses (both generalist, such as those offered at the ISAI, and specialist, such as hotel management and marketing for tourism), there has so far been little contact or exchange of ideas among the English language teachers in these institutions.

A preliminary definition of ETP could divide its domains into four areas: face-to-face customer service, back-office routine, line management, and strategic planning. Face-to-face customer service would include all the language required to deal with routine bookings of items such as hotel rooms, car hire, and air tickets, either at a service desk or over the telephone. Tourist guides' and tour leaders' interaction with clients would also fall into this first category. From a teaching point of view, correspondence and writing of all kinds can be regarded as back-office tasks in that they do not require direct contact with the client. In the area of management, elements such as marketing, staff recruitment, negotiation, and budgets might be viewed as line management, whereas outside influences on the tourism market, such as ecological awareness, the Internet, or the globalization of travel brand names, would fall under the category of strategic planning. Many of the management elements are common to many areas of business English as well.

This characterization of ETP offers a broad prioritized view of appropriate situations for inclusion at different stages in the degree course—the customer service language with its emphasis on polite register is more appropriate for the earlier part of the course, but the strategic management elements are more appropriate at the end. This classification also corresponds to some extent to the other content areas of the degree. Thus, as students grow in their knowledge of tourism, they can more easily understand more difficult English and the more advanced contexts within which it functions.

◈ DESCRIPTION

ETP at the ISAI

Students studying for a degree in tourism at the ISAI attend 20–23 hours of classes per week during each of their 3 (pass degree) or 4 (honors degree) years of study. The courses are broadly based with the objective of equipping students to work in hotels, airlines, tour operations, travel agencies, or government organizations. Thus, during their 3 or 4 years in our school, students take courses in economics, accounting, Portuguese law related to tourism, geography, communication, marketing, statistics, hotel studies, travel agency studies, general history, and art history as well as languages. All courses, except languages, are taught in Portuguese by Portuguese staff members.

The students at the ISAI study two foreign languages, each 4 hours a week, for a total of 8 hours a week over 3 years. Foreign languages are not part of the fourth-year curriculum. All students must take English as their first foreign language and select either French or German as their second. English classes vary in size from 20 to 35 students. However, actual class attendance varies over the year and may be somewhat lower. As classes for all languages are organized by year, levels within a class, especially in the first year, can span a very broad range. At the time of writing, two experienced teachers shared the teaching workload.

When I began teaching in this program in autumn 1996, our target was to enable our third-year students to reach a level similar to that of the Certificate of Proficiency in English (CPE) exam, with the Certificate in Advanced English (CAE) examination as the second-year target and the Cambridge First Certificate in English (FCE) examination as the first-year target. It quickly became obvious, however, that our goals required some revision. Essentially, the CPE is more academic than necessary for a professional course, and its approach makes it difficult to adapt to an ESP environment. The choice was governed by a perception that an external reference point helps in conceptualizing standards and a belief that students whose degree includes such a major language element should reach a high level of competence.

Currently, we use the FCE as a first-year benchmark and use the CAE as a general target for the second and third years. Within the 5-point Association of Language Testers in Europe (ALTE) scale, the CAE is a Level 4 ("competent user") examination. According to the *ALTE Handbook*, "Examinations at Level Four may be used as proof of the level of language necessary to work at a managerial or professional level or to follow a course of academic study at university level" (UCLES, 1998, p. 6). Although teachers have no control over levels in first year, semester exams and final exams at the end of the first year are written with the FCE level in mind to ensure that those who enter the second year have achieved a certain level of general language competence as well as some knowledge of ETP.

Our overall emphasis has moved toward developing specific oral communicative competence and a knowledge of specific tourism vocabulary. Within this panorama, the first year is less particular to tourism and more grammar oriented (see Table 1). It is essentially a combination of classic FCE elements and some tourism elements. The tourism content is confined largely to the transactional language mentioned as the first area in the definition of ETP given in the Context section. The second and third years are increasingly specific to English for tourism, emphasizing more technical language as well as improved performance on transactional tasks.

◈ DISTINGUISHING FEATURES

Goal-Directed Learning

Our program is guided by three educational objectives based on how our students will likely use English once they graduate.

TABLE 1. OUTLINE OF CONTENT BY YEAR

Year and Level Benchmark	Goal	Sample Tasks and Materials
1. First Certificate in English (FCE)	Ensure foundational English for general purposes Introduce basic ETP	• Complete FCE language exercises • Introduce presentations • Introduce descriptive language: hotel, food, car hire, tour/travel advice, telephone calls, reception
2. Certificate in Advanced English (CAE)	Introduce situated ETP	• Teach general ETP speech and texts: simple tourist brochures, situational language and complaints, letters and reports, tour presentations
3. CAE and beyond	Improve ETP language and thinking skills	• Study complex ETP discourse • Study complex tourist brochures, curriculum vitae, applications, reports, financial issues, strategic planning

If actually employed in tourism, most students pursuing a degree in the field hope eventually to achieve a management position of some kind. However, tourism tends to demand that aspirants for higher positions start at the bottom, dealing with clients at the service level. Therefore, one of the objectives of our English instruction is to ensure that students can deal with routine face-to-face communication related to hotel, automobile, and airline reservations at tourist office information counters and routine voice-to-voice communication over the telephone. The program places a strong emphasis on these tasks, particularly in the first 2 years of the course.

The second objective is to ensure that graduates can use language in a more analytical way by investigating topics such as niche markets, finance, and ecotourism as well as future problems facing the industry. The purpose is to equip graduates with the language skills needed for the English they will encounter at higher level positions in the industry.

The third objective of our program is driven by a local peculiarity. Many students want to pass the exam they must take to become officially accredited tourist guides, and a significant portion of the exam involves speaking about Portuguese monuments and attractions. Portugal has a rich architectural heritage, reflecting its years as an important world power, with the result that tour guides spend much time talking about the historic and architectural aspects of various national monuments. Teaching students to speak intelligently about these in English is the third objective of our program.

Relevant Materials

Although first-year classes tend to focus on grammar, second- and third-year courses are directly related to the language of tourism and its proper application. The key questions we ask before using or adapting any material for class are whether it has any direct relevance to tourism and whether it is related to the content in other subject areas. We have been unable to find suitable textbooks for second- and third-year students, and, as a result, we are constantly creating or adapting material so that it is relevant to the course of studies as a whole. When students study Portuguese monuments or various epochs in their art history courses, we discuss the elements of Baroque and Gothic architecture in our ETP class and ask students to give oral presentations about the major monuments in Portugal. Similarly, we ask students to study marketing material from resorts in the United States and have them produce their own versions for attractions in Portugal.

Learning Activities in Reinforcing Sequences

Two important changes took place from 1997 to 2000. The first was an increasing emphasis on student presentations of various kinds, ranging from 1- to 2-minute talks about something just covered in class to 10- to 15-minute presentations of project work. The second type of presentation results from a conscious attempt to link activities into reinforcing sequences as far as possible. For example, letter writing might consist of a sequence in which a letter of inquiry leads to a booking, which leads to a complaint, which in turn generates a response to the complaint or a presentation, followed up by a report. Teachers can link oral and written tasks in a similar manner by having students follow up a task involving a telephone reservation with a written confirmation.

When preparing presentations for assessment in the English class, students are encouraged to use projects that they have already completed for content area courses as the basis for their work. For example, students who have already spent a considerable amount of time preparing a project on tourism opportunities in a certain region of Portugal might base their presentations for English class on the same material. The objective is not only to develop their English but also to reinforce their learning in content classes.

Testing and Evaluation

Grades given in February and July are averaged for a final grade based on class work (weighted as one third of the grade) and examinations (weighted as two thirds of the grade). The final grade is expressed as a whole number out of 20 points: 10 means pass, 12–13 means good, 14–15 means very good, and 16–17 means excellent. Grades of 18 or higher are rarely if ever awarded. Students who fail in February must take an oral exam in July in addition to the written exam.

Testing and evaluation changed considerably from 1997 to 2000. The main trend was to increase the weighting for oral competence within the assessment (classroom) element through the use of presentations for evaluation. Although the first-year written tests are still broadly based on FCE-style tasks, the second- and third-year tests now involve tasks more directly related to what tourism professionals encounter in the working world. The third-year tests are the culmination of the students' language studies and mark the first point at which they may graduate to become tourism professionals.

First-year exams are largely grammar based, using cloze exercises, error correction, and sentence rewriting tests of the type found in the UCLES FCE exams. Students complete a single, relatively simple, formal writing task, as illustrated by the following two items involving report writing and formal letter writing, respectively:

> Your local tourist information office is putting together a leaflet for visitors. You have been asked to write a report on shopping facilities. Write your report of approximately 200 words.

> Write a letter of complaint to Blueskies Holidays, 123 East Road, London 3RC 4CE. You have returned from a holiday in Denmark and are very disappointed because the hotel was not as you expected. Instead of a room with a bathroom, a balcony, and a sea view, you were given a small room with no bathroom or toilet and no balcony, and the only view was the car park. Use your real name and address.

The second-year test uses grammar exercises of the type found in both the FCE and CAE exams. Responding to the exam questions requires specialized, tourism-related vocabulary and a general awareness of tourism issues. The first example below requires specialized architectural or art history vocabulary, and the second requires a wide range of vocabulary covering geography and climate as well hotels and transport.

> Write a brief description of a monument you know well. Your description should include both the architecture and the history of the monument.

Write a general-purpose tourist information sheet for international tourists visiting Portugal for the first time. Use the following headings: Vital Statistics, Climate/When to Go, Getting There, Getting Around, Accommodations, Appropriate Dress, Safety/Security, Further Information. If you are not certain of some information, make a reasonable guess.

Third-year students have already proven their fundamental language skills, so the exams focus far less on explicit grammar. Figure 1 shows a third-year exam that contains no explicit grammar questions. The exam requires the student to demonstrate sufficient language skills to confidently discuss significant, real issues in the tourism industry using appropriate language at a proper level of sophistication. Section 1 offers a choice of topical essay subjects, all of which require students to draw on their subject-matter knowledge from the degree as a whole. Sections 2 and 3 are based on a 500- to 600-word case study that details the implementation of an incentive scheme with undesired effects. Section 2 consists of simple comprehension questions, and to complete Section 3 students must understand the text very clearly and be able to analyze the problem in a business context.

Second- and third-year students who must take an oral exam prepare a short talk about some aspect of tourism in Portugal. This talk typically involves describing

---

SECTION 1 (30 points/6 *valores*
[points given in the Portuguese grading system])

Write an essay of 180–300 words about the difficulties facing Portuguese tourism in one of the following areas:

- agritourism
- tourism in the Algarve
- Porto as a weekend destination.

For Sections 2 and 3, read the case study on the next page and answer the following questions.

SECTION 2 (10 points/2 *valores*)

1. a. What is Cindy's job?

   b. Where is the hotel?

   c. How much is the bonus that she introduced worth to a winning employee?

2. What were the complaints that Cindy was trying to solve with her new bonus scheme?

3. What new complaints began after the introduction of the new bonus scheme?

4. What problems have the other managers in the hotel noticed as a result of the new bonus scheme?

SECTION 3 (30 points/6 *valores*)

Explain what is wrong with the incentive system that Cindy set up, and say what (if anything) can be done to improve the situation. (Do not write more than 150 words.)

FIGURE 1. Third-Year Examination

a tourist attraction of some kind, which leads to an open-ended discussion. For students taking the oral exam, the score is averaged with that on the written test to produce a final grade.

◈ PRACTICAL IDEAS

Many of the activities described below are adaptable to a broad range of intermediate to advanced students; they can be repeated with increasing levels of sophistication as the students progress from Year 1 to Year 3 of the course. As all of the activities described here are directly related to workplace needs, motivation to do them well tends to be very high.

Supplement Textbooks With Authentic Materials

To date, we have not found any ETP textbook entirely satisfactory as a primary text for our context. However, new books are beginning to appear on the market at an increasing rate, so the chances of finding a suitable one have improved. Two supplementary books, *Test Your Business English: Hotel and Catering* (Pohl, 1996) and *Tourism* (McBurney, 1996), have proved somewhat useful. Both also lend themselves to self-study, as they include answer keys.

Besides textbooks, vast amounts of material can be used in ETP classes: travel magazines, travel supplements, and advertisements in newspapers; tourism textbooks for native speakers; travel brochures; first language (L1) and second language (L2) encyclopedias; travel guides; and materials found on the Internet.

Brochures, the basis for many activities, are worth collecting. We tend to have more brochures from the United States than from other countries thanks to the customer response forms in U.S. tourism and travel magazines, such as *National Geographic Traveler*, that enable one to order a large number of diverse brochures by checking them off on a list. *Wanderlust* magazine, published in the United Kingdom, is another favorite: It includes useful country files that develop the students' content knowledge and reading skills and serve as models for presenting factual information.

Include a Variety of Workplace-Related Activities

City Guide Leaflet and Tourist Office Role Play

First-year students in groups produce a small guide to Porto, the city where the school is located. To compile the guide, they write short (80- to 150-word) selections under a number of headings—usually transport, the sights, shopping, nightlife in general, and bars and discos—for a target reader: a young tourist visiting the city. We selected these headings because many of the students are new to the city, and these aspects may well be precisely those the students have come to know in their first few months there, when they are in a sense still tourists themselves.

The activity is carried out in three stages:

1. Pairs or individuals are each assigned one of the headings and write a short description suitable for a leaflet aimed at tourists of their own age. At the end of the period, the teacher collects the descriptions.

2. The teacher pastes together a leaflet containing all the headings and descriptions. In the next class, the teacher reviews any necessary language and gives the leaflet to pairs of students, who produce their own version of the leaflet by editing it and correcting the mistakes.

3. Students are divided into tourist office staff and tourists. Each tourist is given a role card with instructions to ask for certain information about the city. The tourist office staff uses the student leaflet plus personal knowledge to answer the questions. A sample role card might read, "You are a student and have just arrived from Switzerland on an Inter Rail pass. You want to know the cheapest way of getting round the city and would like to know where to find nice bars that are not too expensive. You also want to visit some galleries and see the most important sights." To vary the role play, teachers use extracts from city guides to other cities or tourist office/tourist games from resource books (e.g., Hadfield, 1984).

Short and Long Presentations

Teachers frequently assign short presentations in class. The initial objective is simply to ensure that all the students overcome their nervousness and actually present something. Student motivation is generally high, and the combination of nervousness and excitement usually means that students are actively engaged throughout the preparation period. Various materials, both L1 and L2, have proved useful in preparing presentations: English language encyclopedias, a 12-volume guide to the UNESCO World Heritage sites published in Portuguese, and a collection of English language tourist brochures and guides of all kinds as well as short reference works on local art history.

Short presentations (2–7 minutes) usually require two consecutive class periods. Students in pairs look at a volume of a reference book and choose a monument, a city, or another tourist attraction, about which they make notes and then give a presentation to their colleagues in the following period. Often students consult brochures and present some aspect of a tourism region or simply introduce a particular hotel or resort. During the preparation, the teacher circulates from one group to another to answer vocabulary queries, suggest various ways of phrasing certain ideas, and give individual attention to grammar questions that frequently arise.

A difficulty is ensuring that the students listen when their colleagues are presenting. One method is to have the students rate all the presentations according to a simple scale using a worksheet. An alternative is to have the students take notes during the presentations and ask a few of them to summarize what a speaker has said.

Ten-minute, independently researched presentations are used for evaluation. As noted in the Distinguishing Features section, a major objective is to familiarize students with speaking about Portuguese monuments and attractions before an audience, as this kind of task features prominently in the examinations that students must take to become licensed tour guides. These presentations are also a means of activating vocabulary related to, say, Baroque architecture by having students describe an attraction such as the Batalha Cathedral. As follow-up, the students usually write an essay in class in which they outline some aspect of their presentation and reflect on what they would do differently in hindsight as well as on what they have learned from the experience.

Brochure Activities

Brochures lend themselves to the learning of descriptive language in context. In brochure activities, students learn to describe resort facilities or landscapes, nature, and outdoor pastimes. The activity takes the following form:

1. Students receive a brochure and

   • list vocabulary/phrases under specific headings

   • make notes about a certain attraction or place

   • prepare and deliver a very short (2- to 3-minute) presentation about the attraction or place

   • give the vocabulary/phrase list to the teacher

2. The teacher edits the lists and combines them into one handout. In a subsequent class, the students write a description of an imaginary attraction, a resort, or a geographical region using some of the new vocabulary. The teacher circulates and provides assistance.

3. In a role play, half of the students act as tour company representatives, and the others act as representatives of the imaginary location they described. The tour company representatives visit the location representatives in search of new attractions to put on their program, and the representatives try to "sell" their region.

Other Role Plays

Role plays as a teaching strategy can be used in many ways. In another role play I have used, the student portraying the travel courier or guide has to help with a problem (e.g., food poisoning, theft, forgotten medicines, problems with rooms). One of the key objectives is to encourage students to use sympathetic language and express the emotion they might normally employ in their L1 in similar circumstances (e.g., *Oh dear, I'm terribly sorry. Are you hurt? What happened?*)

Letters of complaint are easily linked to this role play. Students study a number of model letters and write their own letter(s) for a particular situation. An effective sequence is to have students first write letters of complaint and then write appropriate letters of response from the point of view of the agency, the tour operator, or the tourist establishment.

Role plays involving negotiation skills are quite interesting and reasonably easy to prepare. The basic scenario involves two parties, each of whom has something the other wants, for example, a hotel in a beach resort that wants to do business with a tour operator to bring tourists from England to Spain or Japan to Hawaii. If negotiation proceeds well, both parties get something close to their ideal solution— a win-win outcome. If the negotiation ends prematurely with, for example, one party walking away and refusing outright to negotiate further, a lose-lose situation occurs. The intermediate solution involves one party getting what it seeks while the second party fails to achieve all of its objectives. In some circumstances this role play may be developed as a complete module.

I tend to use negotiation role plays close to Christmas, when I can illustrate some basic strategies by referring to the students' experience. Specifically, a trip to a major store at Christmastime reveals toys piled up close to the entrance to catch children's attention. One can then observe a classic negotiation scenario as parents try to navigate their children past the toys without being forced to buy something. Parents promise the children that Santa Claus will bring them something special if they behave, that is, stop crying and stop demanding the expensive toy. Worst-case scenarios can be painted: "Santa Claus only brings gifts to good little children." The arrangement may be sealed with a "sweetener," that is, "if you are really good, then maybe we'll get some ice cream on our way home." When we discuss negotiation in this context, the students immediately grasp the concept.

◈ CONCLUSION

This program changed substantially over the 3 years described in this chapter (1997–2000) and will continue to develop as new elements are tested and either added to the program or discarded. In particular, we will be looking at publications aimed at native speakers (e.g., Hinkin, 1995; State Tourism Training Agency, 1997; Syratt, 1992, 1995) as well as for sources of new material. Although ideas such as teaching part of a content course in English or one of the other foreign languages might be effective, such an idea is unlikely to be implemented soon. We will, however, continue to relate activities in the English program to the content of other areas. In particular, we would like to develop more material related to local geography and art history in cooperation with the content teachers in these areas.

The program will probably also increase its use of the Internet. Initial explorations in which students created some pages related to tourism proved interesting; however, it is too early to say precisely how this area of the curriculum will develop. The formulaic nature of certain repetitive tasks, such as teaching the basics of formal letter writing and giving examples for curricula vitae and reports, makes them potential candidates for an Internet approach. To avoid duplication of effort, teachers in a number of tourism schools might cooperatively develop on-line materials that all could share whenever their learners' needs overlapped. Internet applications that use lowest common denominator Internet technologies, such as Common Gateway Interface (CGI) or JavaScript, have the advantages of being largely platform independent; another advantage is that suitable authoring tools are becoming easier to use and less expensive.

An ambitious but unlikely step within the Portuguese educational system at present would be to restructure language teaching to take better advantage of each student's prior knowledge. This would involve moving from a year-based system to a level- or module-based system, with credits given for what a student has already learned. In this context, some elements could become largely self-study modules with the teacher becoming a mentor or tutor.

More generally, training in languages (including English) will become ever more important within the tourism industry, and many of the language skills required are identical across languages as tourists worldwide have essentially the same set of needs and the tourism industry the same set of products. A systematic approach to defining these language requirements would be beneficial for all involved.

◈ CONTRIBUTOR

Simon Magennis began teaching EFL in 1986 and has taught in Italy, Spain, Portugal, and Ireland in universities, private language schools, and summer schools. He has taught and designed a wide variety of courses on subjects ranging from business English, English for tourism, English for pharmacy students, and composition, to U.S. culture, history, and institutions. He has also assisted students in preparing for the UCLES exams. He worked as a lecturer in English for tourism and for translation at the ISAI in Porto, Portugal, from 1996 to 2000. Originally trained in natural sciences, with experience as a geologist, he has a strong interest in the Internet and technology related to learning. He is now based in Ireland, where he works in information technology training for InHouse Training International.

CHAPTER 5

An ESP Program for Students of Shipbuilding

Elena López Torres and María Dolores Perea Barberá

◈ INTRODUCTION

In this chapter, we describe an ESP program for the learning of technical English for shipbuilding in the College of Naval Architecture and Marine Engineering at the University of Cádiz, Spain, as well as in the surrounding shipbuilding companies. We also discuss the relationship between the university and the naval industry, from which university ESP teachers can gain an understanding of the professional world they serve. The program is organized around a set of technical topics and objectives derived from a needs analysis and set in the syllabus. These topics and objectives have served as the basis for a textbook and a multimedia application for this line of ESP.

◈ CONTEXT

Cádiz: Location and Maritime Tradition

Cádiz is a Spanish province located in the southernmost part of Europe, close to the Strait of Gibraltar. It has a long maritime and ship construction tradition. Since the discovery of America in the 15th century, most of its working activity has been related to maritime commerce, shipbuilding, the navy, and the fishing industry. In recent years, offshore drilling platforms for the exploitation of energy resources, aquaculture equipment, and ships built for leisure activities, ranging from small sailing boats to yachts and ferries, have constituted an increasingly important sector of the local industry.

Nowadays, some of the most important companies in Spain related to marine technology are Izar, with shipyards in Puerto Real, Cádiz, and San Fernando, and Dragados Offshore, all located in the Bay of Cádiz. The facilities of Astillero de Puerto Real, one of the largest and most advanced shipyards in Spain, are spread over an area of 1 million m². The building dry dock is 500 m long by 100 m wide, and it is served by two gantry cranes that together can lift more than 1,000 tons. The yard can build oil tankers, big container ships, and bulk carriers as well as cruise ships and large passenger ferries. Astillero de Cádiz, with its large-scale facilities and close proximity to the Puerto Real yard, can deal with huge conversions and specializes in ship repairs. Cádiz has three dry and floating docks and more than 2 miles of piers for outfitting. The shipyard in San Fernando specializes in the construction and outfitting of warships; since the early 1990s, it has been developing its own technology to design and build aluminum high-speed ferries. Dragados Offshore specializes in offshore drilling platforms. Furthermore, more than 50 small and medium-sized enterprises are devoted to the manufacture of outfitting equipment for ships, the building and repairing of propellers, metalwork, and other activities that support the local shipyards.

The naval industry in Cádiz is clearly a very important aspect of the local economy. In addition, the international nature of shipbuilding makes English the compulsory language of communication and accounts for the demand of highly specialized ESP courses.

Academic Environment

Founded in 1978, the University of Cádiz, which has about 20,000 students and 1,600 teachers, offers degree programs in medicine, law, history, philosophy, linguistics, economics, business administration, chemistry, chemical engineering, industrial engineering, industrial management, naval engineering, and nautical and marine sciences. The university has campuses in Cádiz, Puerto Real, Jerez, and Algeciras. One of the newest and best equipped centers for higher education in Cádiz is the Andalusian Center for Higher Marine Studies. Three faculties, or colleges, concerned with marine studies are housed there: the Faculty of Nautical Studies, with degree programs in naval communication and radioelectronics, sea navigation, and marine engineering; the Faculty of Marine Sciences, with degree programs in aquaculture, marine ecosystems, and marine environmental studies; and the College of Naval Architecture and Marine Engineering, which offers degrees in naval architecture, marine propulsion, and offshore industry. Most of the lecturers in this college either worked previously in the maritime industry or remain employed in the maritime industry while teaching part-time at the university.

The ESP program *Inglés técnico naval* (English for Shipbuilding), described below, is part of the curriculum for a technical degree in naval architecture and marine engineering. A textbook and a multimedia application serve as tools for the learning of this special type of English. Furthermore, this ESP program has been designed for use in professional contexts within the maritime industry.

Professional Context

Since this ESP program began more than 10 years ago, we as educators of prospective naval architects and marine engineers have felt a need for and interest in close cooperation with local industries. On the other hand, the international character of most orders and contracts has meant that shipyards and related enterprises have regarded English as important, as reflected in the fact that the main shipyards have a permanent scheme of general English courses concentrating on spoken English, mainly conversation.

Some of the most important documents used for the building of merchant ships in Spain are conventions adopted by international organizations. The most important of these, regarding maritime legislation, is the International Maritime Organisation, an agency of the United Nations, which has 155 member states. The second block of official documents of compulsory application in shipyard practice comes from classification societies, which guarantee the initial and continuing inspection of ships. These societies publish up-to-date standards, referred to as rules and regulations, for the construction of ships and their machinery. The third type of official documentation is contracts and accompanying documents, such as ship specifications, plans, and procedures. Except for the rules and regulations, which are not available in Spanish, all other documents are published in both Spanish and English. In contracts signed in English with a foreign shipowner, the English version is legally binding as far as contract documentation is concerned.

Our research and cooperative projects with the maritime industry have focused mainly on the shipyards in Cádiz and Puerto Real. The projects have included technical translation and research toward the compilation of databases for computer-assisted translation as well as teaching English for shipbuilding, a job that teachers of general English could not assume because of its highly specialized nature. In recent years we have monitored several intensive 3-month ESP courses with 10–20 students per class and a strong emphasis on conversation about ship construction topics. In addition, from January to June 1999, the Astillero de Puerto Real shipyard developed a new approach to ESP learning that demanded a semiautonomous course based on a multimedia application developed by our research group. This approach was designed to allow working professionals to fit their studies more easily into their busy work schedule.

Student Profile

Two categories of students participate in this program: university students and working professionals. Several features of the students at the College of Naval Architecture and Marine Engineering stand out:

- The minimum age of the students is 17; the average is 19–20.

- Although more women have chosen technical degrees in the past decade, the majority (about 80%) of the students are male.

- Most of the students come from the provinces of Cádiz and Seville, which combine a long maritime tradition with up-to-date technology for ship construction. Around 20% come from other parts of Andalusia, Castile, Catalonia, or the Canary Islands. Another group of students from the Maghreb, particularly Morocco, is unusual in that Spanish is their third language, Arabic and French usually being their first and second languages respectively, and they learn EFL in Spain. Their level of English is usually beginning or false beginning, and taking the ESP course requires considerable effort on their part.

- Because English for Shipbuilding is a first-year course, the students must learn the concepts and techniques of shipbuilding in Spanish from their lectures in naval architecture while learning English terminology and features of technical discourse from their ESP teachers. This situation has conditioned, to a great degree, the topic-based approach in the ESP program, as discussed below.

- Their level of general English tends to be intermediate or preintermediate. Similar to technical students elsewhere, they can interpret diagrams, graphs, tables, and drawings but have difficulty employing proper grammar, providing accurate translations, and summarizing lengthy material into shorter, simpler forms.

The second group of students, those in the workplace,

- are adult learners from various backgrounds with varying levels of competence in English

- are highly motivated because they are aware of the importance of English in their workplace

- come from various fields of expertise and use their previous knowledge of the subject matter to help them understand the language

- generally have more highly developed skills in autonomous learning than do students in the first group

◈ DESCRIPTION

English for Shipbuilding is a nine-credit, first-year compulsory course for students taking degree courses at the College of Naval Architecture at the University of Cádiz. The course extends over two semesters, and the number of students registered is roughly 150, divided into two groups. Theory classes are allotted 4.5 credits and meet once a week in groups of 75–100 for a 90-minute lecture. For the 4.5 credits of practical classes, the students are divided into groups of 20 that meet once a week, from November to May, for a 2-hour lesson in the language lab. These classes concentrate on social aspects of language and general English.

The Syllabus

Design

A commonly accepted tenet in ESP theory (Robinson, 1991) is that the analysis of student needs is one of the first activities in the design of a course. In creating the syllabus for English for Shipbuilding, we broadly followed the relevant categories established by Munby (1978) in *Communicative Syllabus Design*—purposive domain, setting, interaction, instrumentality, dialect, target level, communicative event, and communicative key—carrying out a series of surveys before delimiting the amount of terminology to include in the course. Our three basic sources of information for the needs analysis were students taking degrees in naval architecture and marine engineering, academic authorities, and professionals. Helpful input from these sources made possible several conclusions regarding the design of the syllabus:

- The domain should mainly be ship construction techniques.

- The students need to read bibliographies in English in an academic environment.

- Learners in the working environment need to read professional texts, mainly rules of classification societies and contracts, and to write specifications, procedures, and reports for the shipowner as well as letters and faxes.

- Learners in the working environment need to communicate orally during business meals, in conversations with classification society surveyors, and during contract negotiations.

- The context for all spoken and written interactions is mostly formal, though informal situations also exist.

- In the majority of cases, interactions take place among people of different nationalities using English as an international language for communication and with many different phonetic peculiarities.

- Our informants clearly preferred learning English as a tool and not as an end in itself.

- Our informants were aware of the importance of learning the terminology necessary to understand and participate in different types of spoken and written discourse.

For several reasons, the syllabus designed for the ESP classes in both the academic and professional contexts is content based:

- Students need to learn and practice the different skills through the specialized language and terminology of the their fields of studies. Learning language by using it as a tool has proved to motivate technical students.

- The large amount of terminology students need to know if they are to cope with various types of texts and documents during degree studies and in their professional lives demands a content-based approach.

- All courses taught at the College of Naval Architecture and Marine Engineering (except English) use Spanish as the language for communication. And first-year students in the ESP course are learning ship construction concepts in Spanish while they learn English terminology. This fact influenced the types of text used for the lectures, which were finally collected in the textbook *Inglés técnico naval* (López, Spiegelberg, & Carrillo, 1998; see below).

- A content-based syllabus increases learners' awareness of the importance of vocabulary to language and language learning.

Another important aspect of syllabus design was determining appropriate categories for organizing the ESP instruction. After considering several options, we decided to base the categories on the headings and subheadings of textbooks used in the College of Naval Architecture and Marine Engineering, which were directly related to ship construction. Once we had outlined the topics, we showed the resulting organizational structure to various subject specialists to obtain comments and recommendations. Only after completing this research could we establish the final syllabus.

Content

The syllabus, designed to be used in both academic and professional contexts, includes a wide variety of topics:

- engineering topics, such as the unit "Principal Materials for Ship Construction," in which the classification of the most commonly used metallic materials, both ferrous and nonferrous, is followed by their properties and different uses, and "Methods of Joining Structural Parts:

Welding," in which texts and exercises center on various types of welding and welding equipment

- the classification and description of merchant ships, as in "Dry Cargo Ships," "Liquid Cargo Ships," and "Auxiliary Ships"

- ship dimensions, as in "Principal Particulars of Vessels" and "Draught Marks and Tonnages"

- the structure of ships, as in "Ship Subdivision," "The Hull Structure," and "Decks"

- the description and organization of a shipyard, as in "Shipyard Practice"

- legal aspects and building standards, as in "Classification Societies" and "Freeboard and Load Lines"

- propelling systems, anchoring, and cargo handling, as in "Propellers," "Anchors and Cables," and "Hatchways and Hatch Covers"

The practical classes taught at the College of Naval Architecture and Marine Engineering focus on English for social contacts. The syllabus consists of units based on functional and communicative approaches to language learning. As classes are small (20 students per group), practice in oral skills is encouraged. Furthermore, because language lab facilities are available, the classes also emphasize listening—the *Cinderella skill* (Nunan, 1997)—and pronunciation practice. The classes cover the following functions:

- *establishing contacts:* Students learn and practice how to greet appropriately in various situations and contexts, how to introduce themselves and others, and how to start a conversation and express their interests.

- *expressing interests and socializing:* Students learn to express likes and dislikes, various degrees of agreement and disagreement, and ways of inviting, accepting, and refusing.

- *traveling:* Students learn how to request and supply travel information using direct and polite questions, make reservations, talk about and ask questions about past events and actions, inquire about exchange rates, ask for information about a country, and give information about Spain.

- *telephone manners:* Students learn to introduce themselves properly on the telephone; connect to an extension; listen for information; and initiate, maintain, and conclude a telephone conversation.

Materials

Inglés técnico naval

The lack of published materials within this ESP field launched us into the world of material design. *Inglés técnico naval* (López et al., 1998) was especially designed to teach naval terminology and was tailored to the students. The book is divided into 21 units on various aspects of ship construction. Each unit is in turn subdivided into two parts: texts in English describing the basic aspects of the topic using normal terminology in simplified sentences (see Figure 1) followed by exercises to practice the new terminology (see Figure 2).

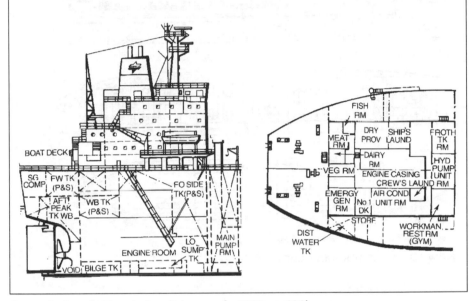

After Peak

The after peak tank is situated at the extreme after end of the ship, immediately abaft the machinery space. This tank is essentially used for water ballast but it can be used for additional fresh water. In the forward side of this tank is the "after-peak bulkhead" which extends from the bottom of the ship to the main deck.

FIGURE 1. Sample Description (López et al., 1998, p. 123)

Each chapter includes authentic material, such as drawings and photographs from the Spanish shipyard archives, forms and certificates from classification societies, and sketches from commercial catalogues, and is introduced by a bilingual vocabulary list of the specialized terms contained in the texts (see Figure 3). The answer key at the end of the text allows students to use it somewhat autonomously. In addition, a glossary contains the Spanish equivalents of the English technical terms included in the book. Accompanying audiotapes contain recordings of the written texts. The different backgrounds of the three authors (a shipmaster, an engineer, and a linguist) ensure that the material is appropriate in both subject matter and language. After issuing three editions of *Inglés técnico naval*, the authors have arranged with the marine technology department of an Argentinian university to coauthor a new version of the book, incorporating new topics related to ships engaged in leisure activities (e.g., sailing boats, yachts, catamarans).

Curso multimedia de inglés naval

The *Curso multimedia de inglés naval* (InterDis Research Group, n.d.) is designed for use in the classroom and for independent study. The English language instruction in

---

E X E R C I S E

EXERCISE A.- Look at this diagram of a cement-bulkcarrier:

ALZADO
UPRIGHT PROJECTION

CUBIERTA SUPERIOR
MAIN DECK

DOBLE FONDO
DOUBLE BOTTOM

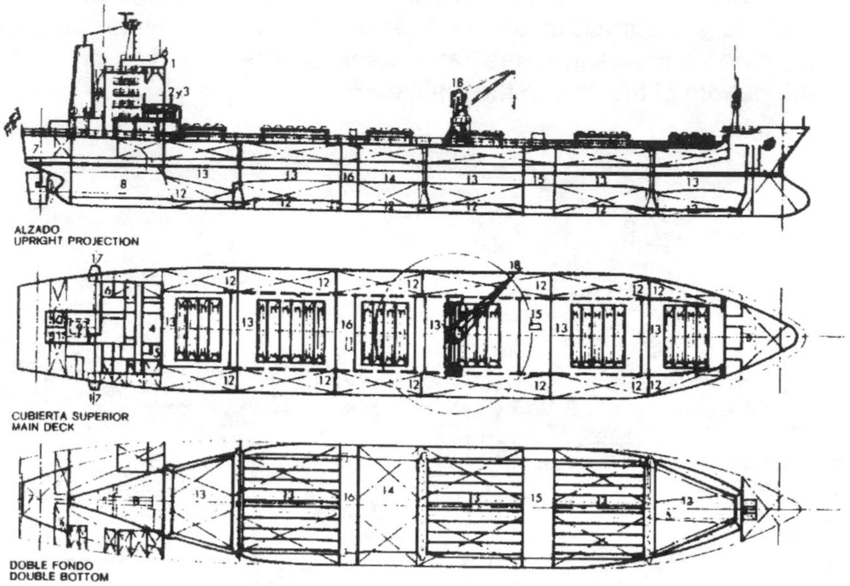

Now say which numbers correspond to the following:

Bridge crane
Accommodation and Services   Engine room
Steering gear   Holds
Wheel house and Chart-room   Ballast tanks

Now, complete this description of the above diagram:

This is the general arrangement of a cement-bulkcarrier. The bridge and the engine room _____ aft. At the extreme bow and stern are two tanks, the _____ _____ and the _____ _____ ; they are used for fresh water or _____ . There is a large number of _____ on the main

127

FIGURE 2. Sample Exercise (López et al., 1998, p. 127)

TEMA 4

FREEBOARD, LOAD LINES AND DRAUGHT MARKS

Vocabulario

Deck line............................ Línea de cubierta
Depth................................. Puntal
Draught Marks .... Escala de calados
Freeboard .... Francobordo
Freeboard deck................ Cubierta de francobordo
Load lines .... Líneas de máxima carga
Load line disc .... Disco de máxima carga
Plimsoll disc...................... Disco Plimsoll

PLIMSOLL DISC

The Plimsoll disc was named after Samuel Plimsoll, a member of the British Parliament who felt the necessity of establishing a mark to indicate the depth to which vessels may be loaded.

It is a disc 12 inches in diameter with a horizontal line 18 inches long passing through the centre, the upper edge of which indicates the maximum depth to which a vessel may be loaded in salt water in summer time.

The letters at each side of the disc are the initial letters of the assigning authority, for example "L-R" means "Lloyd's Register of Shipping"

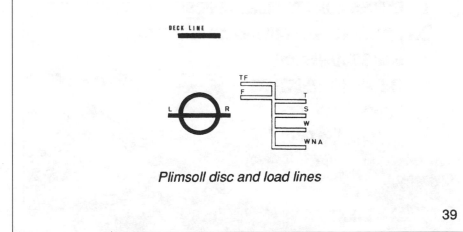

Plimsoll disc and load lines

39

FIGURE 3. Sample Vocabulary List (López et al., 1998, p. 39)

each lesson is useful not only for university students but also for technical staff in Spanish shipyards.

The multimedia application is made up of 10 units. Units 1–9 deal with various topics in ship construction, such as expressions to indicate positions on board and outside the ship, principal particulars of vessels, ship subdivisions, the hull structure, decks, the shipyard, welding techniques, hatches, and various types of merchant vessels, and Unit 10 is an English-Spanish dictionary of technical terms (see Figure 4). Each unit is linked to the contents menu.

When a user enters a unit, an image directly related to the audio and to the written text appears on the screen. The written text is shown only after students listen to it while looking at the visuals on their computer screen. Students control the volume by moving the sound bar (see Figure 5). Terms with a higher level of difficulty appear in a different color in the written texts. When the user clicks the mouse on them, a pop-up window with the translation into Spanish appears (see Figure 6). Each unit is followed by a series of exercises designed to give practice with the new vocabulary. The various types of exercises all enable students to check their own answers (see Figure 7). The number of images used (about 150) responds to the need for graphic support to describe ship construction concepts and were designed to be self-explanatory in order to support the listening comprehension practice.

The dictionary included in Unit 10 is the result of research carried out in the field of English and Spanish shipbuilding terminology and translation equivalents,

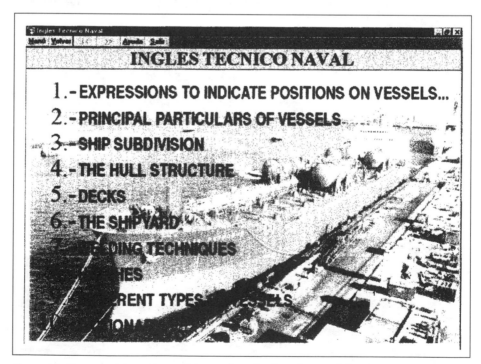

FIGURE 4. Contents Menu From *Curso multimedia de inglés naval* (InterDis Research Group, n.d.)

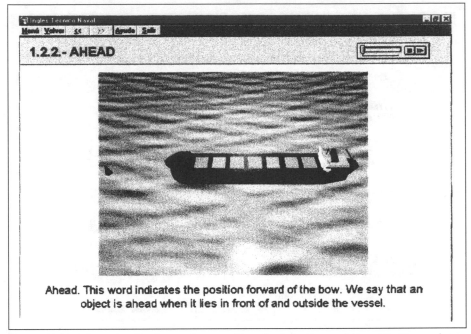

FIGURE 5. Unit Introduction From *Curso multimedia de inglés naval* (InterDis Research Group, n.d.)

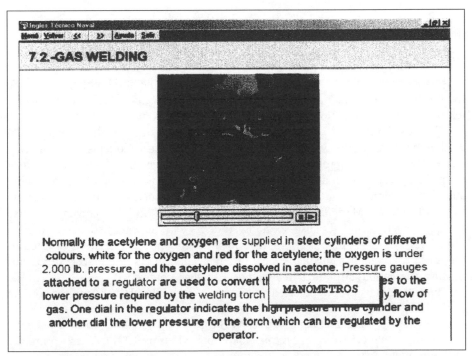

FIGURE 6. Pop-up Window From *Curso multimedia de inglés naval* (InterDis Research Group, n.d.)

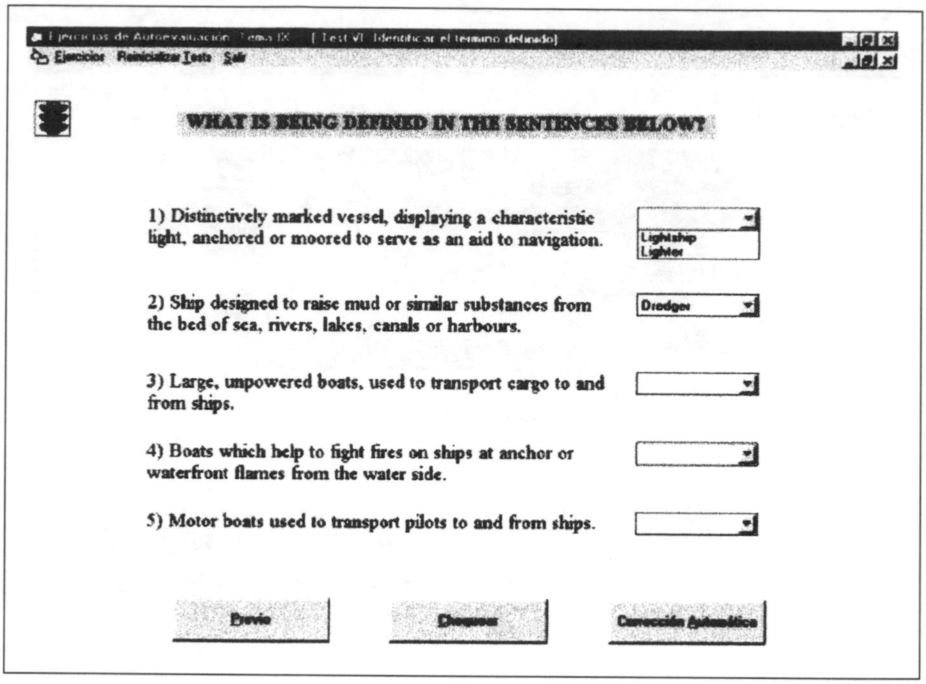

FIGURE 7. Self-Evaluation From *Curso multimedia de inglés naval* (InterDis Research Group, n.d.)

published in *Inglés técnico naval* (López et al., 1998), which served as the basis for the multimedia application.

ESP and Ship Construction Textbooks

As previously mentioned, materials directly related to the shipbuilding world are hard to find. Other materials we have used include ESP textbooks, ship construction textbooks, technical dictionaries, realia, and Internet resources.

The text *English for Nautical Students* (Baker, 1979) introduces the language used in the maritime world, with the aim of helping nautical students communicate effectively in their discipline. Although some units may be of interest to naval architects, most of them are clearly intended for the navigating crew. Despite being aimed at students of nautical colleges and naval establishments, some chapters of the book *English for Maritime Studies* (Blakey, 1987)—an English language course designed for nonnative-English-speaking students with an intermediate knowledge of general English—are also appropriate for our students. The sections devoted to the study of grammar highlight the most relevant aspects and, most important, practice the various structures in a technical English context.

From the literature, we selected several basic textbooks with an emphasis on the engineering fundamentals of naval architecture: *Introduction to Naval Architecture* (Tupper, 1996); *Merchant Ship Construction* (Pursey, 1998); *Ship Construction* (Eyres, 1990); and *Ship Construction Sketches and Notes* (Kemp, 1997). These primarily undergraduate texts are aimed at students in nautical schools and in professional courses in shipbuilding. The authors offer concise overviews of the technical aspects of ship design, which complement our purposes and may be used by our learners for supplementary reading.

Technical Dictionaries

We encourage students to use specialized dictionaries to solve problems of terminology. In English for Shipbuilding, which students usually take during their first year at the university, they receive some information on the various dictionaries available and, afterward, guidelines on how to make the best use of them. Among the recommended dictionaries are *Diccionario técnico marítimo* (Suárez Gil, 1983), *Nuevo diccionario politécnico de las lenguas española e inglesa* (Beigbeder, 1997), and *Diccionario náutico: Inglés-español. Español-inglés* (Malagón Ortuondo, 1998). Maritime dictionaries cover the greatest part of the vocabulary of sea-related domains. As shipbuilding is a subset of the maritime world, maritime dictionaries are bound to include terms related to the ship. However, for terms specifically related to the design of the ship—dimensions and forms or structure, for instance—a more specialized dictionary may be more accurate.

Several class exercises are aimed at developing the students' abilities to solve real terminological questions. In their second year, students who register for *Inglés técnico naval II* (English for Marine Engineering), a six-credit optional course focusing on the terminology and documentation of marine engineering, must hand in a translation exercise that substantially influences their final grades. This requires heavy use of technical dictionaries as well.

Realia

Collecting authentic materials was supremely important to us in the initial stages of this program. We obtained catalogues, sketches and drawings, photographs—some later included in the textbook—terminological glossaries, documents, and other authentic materials from several shipbuilding companies, particularly Izar and Astillero de Puerto Real. We also use videocassette recordings on various related topics, such as the E3 ship project, Spanish shipyards, or welding techniques, to support class activities. Finally, professionals from surrounding companies lecture by invitation on practical aspects of the work in the shipyard, such as the duties of students under work experience contracts, the organization of shipyard departments, and the role of a project manager. (Some students in the Naval Architecture College at the University of Cádiz work for shipbuilding companies to gain experience and further training in their final university years.)

Internet Resources

The Internet has proved to be a source of authentic materials for class development as well as a source of information on shipbuilding for our undergraduate students. Several sites are relevant to our syllabus, among them the Web pages created by various shipbuilders' and naval architects' associations, companies, and legal bodies, mainly classification societies.

Throughout the course, students wishing to improve their final grades have the option of handing in a paper on one of a series of topics provided by the teacher. Apart from the usual sources of information, namely, books and magazines in the library, students receive some guidelines for using the Internet. In our experience, students greatly enjoy navigating the Internet, finding lots of relevant information for their papers while practicing their English. We have also used the Internet to design exercises to build vocabulary. Although the exercises per se are not innovative, students have shown considerable interest in exercises based on texts that contain the most up-to-date information.

◈ DISTINGUISHING FEATURES

InterDis Research Group

One of the most distinctive features of this ESP program is the fact that, though solidly rooted in academia, the university faculty maintains very close contact with industry to identify genuine industrial needs. This contact exists partly because we are members of an interdisciplinary research group with a solid professional relationship with companies, demonstrated by its more than 40 cooperative projects over 11 years. The InterDis research group is made up of researchers from the fields of engineering, computer science, and linguistics. This diversity of background enhances mutual support and engenders an original and enriching approach to research. The group has developed several multimedia applications: a multimedia course on welding techniques and welding inspection, a program for the management of health risks at the workplace, a multimedia application for the selection of welders by prospective employers, and the above-mentioned *Curso multimedia de inglés naval* (InterDis Research Group, n.d.). Some of these are being successfully used outside the university by various enterprises.

Self-Access Multimedia Course in a Professional Setting

As part of a training scheme partially funded by the European Union, professionals working at the Izar factory in Puerto Real were offered a multimedia course on the terminology of shipbuilding. The course was not mandatory, and interested workers were free to register. Simultaneously, native-English-speaking teachers working for a private language school offered a course in general English. In this company, various departments—from the procurement department to the projects and technical division—need to use English for international communication. Personnel need to read and understand specifications, classification society rules, contracts, procedures and other types of documents in English with their corresponding terminology.

Initially, 30 people registered for English for Shipbuilding. The multimedia application was installed on-line in their workplaces, and each student received at least four 1-hour tutorials during the course. The students could also contact teachers via telephone or e-mail when they had questions. As expected, a test revealed that the students came from different backgrounds and had varying levels of English proficiency. Because the course deals with specific shipbuilding concepts, knowledge of the subject matter was also found to be useful. In an introductory class, the students were shown how to navigate through the application and received some guidelines for self-access study. These guidelines, which were based on the design of the multimedia course (see below) included warmup activities (e.g., "Before starting the unit, write down all the English words you can think of related to the topic"; "Look at the picture and write down words or sentences you would expect to hear when talking about it"); listening comprehension, reading, and dictation exercises (e.g., "Listen and tick any word you hear which you have previously guessed and written down," "Move on to the next screen. Listen to the text again while you read it. Note any new words and their meaning"; "Listen and write down what you hear. Use the text on the screen to correct your dictation"); and others. The students were also asked to define their objectives. All mentioned that their reason for taking the course was to improve their proficiency in work-related English and to learn ship construction terminology.

After 5 months of instruction, the students took a test. We established three levels of difficulty, and each student could try any level. Level I, which comprised Units 1–3 of the self-access multimedia course, required the translation of simple sentences from English into Spanish, a listening task (fill-in-the blank), and the translation of a list of vocabulary from Spanish into English. In Level II (Units 4–6), the students were asked to answer questions such as "Do all vessels have the same number of decks? Explain your answer," translate a short text from English into Spanish, give the translation equivalents in English of a list of terms in Spanish, unscramble the paragraphs of a technical text, and complete a dictation exercise. Level III (Units 7–9) included a dictation exercise, translation from Spanish into English, new vocabulary questions, a more sophisticated reading comprehension exercise, and a composition. Around 80% of the students chose the Level I test. Of the remaining 20%, only two of them took the Level III test, and the others chose Level II. All of the students passed their tests with good marks.

For the most part, students, teachers, and managers at Astillero de Puerto Real found the experience quite successful. A survey carried out after the course was completed yielded some interesting conclusions:

- Only one of five students managed to accomplish the learning pace they had set for themselves at the beginning. The rest complained that their usual work activities prevented them from dedicating more time to the course.

- Around 50% of the students had followed the study guidelines suggested to them in the introductory class. All of them, on the other hand, had made extensive use of their own learning techniques and the auxiliary tools that accompanied the multimedia course, namely the textbook *Inglés técnico naval* (López et al., 1998) and the set of audiotapes.

- The main advantage seen in this type of self-access learning approach was its flexibility and adaptability. All the students also valued the tutorials and the opportunities to focus on their individual learning problems.

Applied Research in Specialized Terminology

As the ESP teachers involved in English for Shipbuilding, we are currently developing our careers both as ESP practitioners and as researchers in the field of bilingual maritime terminology and technical translation. Our terminological research is aimed at the creation of terminological data banks and the development of English-Spanish/Spanish-English databases for computer-assisted translation programs, such as IBM TranslationManager (Version 1). The primary source of information for the projects is a corpus of specialized texts and documents in the field of shipbuilding.

This bilingual corpus of parallel texts will provide a reliable source for the identification and retrieval of terms and translation equivalents. Using this database, linguists will be able to study multiword units, degrees of equivalence, collocations, and textual variants. What is innovative in this work is the ability to study terms in context (Pearson, 1998). Traditional terminology, prescriptive and onomastic, is influenced by the new research methods. Modern terminology assumes a rather descriptive nature and accepts, in a way, the semasiological approach that has always characterized lexicographical work. In ESP, a better understanding of the terminology of a field helps considerably in the production of teaching materials and reliable glossaries. Activities designed to promote the acquisition of the terminology of shipbuilding are thus based on a thorough analysis of the terminology in authentic texts.

◈ PRACTICAL IDEAS

Use a Content-Based Approach

Experience in the ESP arena leads us to an instrumental use of English and a content-based approach to language learning. Using English as a tool and content as an end has brought about good results. Because content is considered important, we have successfully used a textbook with content in the students' area of study, in this case ship construction, and with the language adapted for nonnative speakers of English—simplified structures and controlled vocabulary being the most relevant features.

Raise Students' Awareness of the Role of English

A problematic aspect of ESP teaching is motivation. Making students aware of the role of English in their field has proved to be effective, for

> in many parts of the world, university students, for example, may not see the value of their ESP course, perhaps because they did not choose to study their specialization or because they can in fact pass their subject examinations without knowledge of English. (Robinson, 1991, p. 82)

The fact that companies use English tests as filters in hiring new personnel demonstrates the status of English in shipbuilding. The English test, commonly a technical translation from English to Spanish, takes place before any other technical exam or interview, so English may be the key to applicants' success in getting their first job. This knowledge has a strong consciousness-raising effect on students of English.

Cooperate Closely With Industry

Sometimes—particularly in Spain—industry professionals harbor a certain mistrust toward embarking on projects with a university. To overcome this, academic faculty should bear in mind that companies are interested not in theoretical work but in practical results. When funding projects, companies are investing money in exchange for a concrete answer to their needs. Objectives should be realistic and well defined from the very beginning, or disappointing results may affect future projects. University teachers must learn how to adapt to the work pace of professionals, understand how difficult it is to establish cooperative ties with them, and patiently try to take full advantage of their knowledge.

Draw on New Technologies and Research

Computer-assisted language learning appears to be well suited to the task of learning technical terms. Multimedia in its present state is particularly helpful to language teachers in the traditionally neglected area of vocabulary teaching (Milton & Jacobs, 1995). The purpose of introducing a multimedia application is to help students move toward autonomy in the learning of specialized languages while tutors act as supervisors and counselors, guiding peer groups and individual students. If ESP practitioners work together with teachers from computer science departments, they can initiate projects to develop multimedia applications for their fields. Furthermore, if the ESP course is part of a university curriculum, research in terminology, methodology, and applied linguistics results in more qualified teaching staff.

◈ CONCLUSION

The link between ESP faculty in the College of Naval Architecture and Marine Engineering and working professionals—both at the shipyard and at the university—has been very productive. Offering the English for Shipbuilding course in academic and professional settings, with different groups of students, has highlighted the versatility of the materials designed for this ESP setting: Both professionals and college students have found the textbook and the multimedia application perfectly suitable. In spite of curriculum and contextual constraints that have influenced course designs (e.g., ESP courses at the university are taught in lecture theaters with a large number of students per group whereas ESP courses in a professional setting are smaller and more individualized), a content-based ESP approach—with an emphasis on vocabulary teaching—has proved appropriate for university students and working professionals in the field of shipbuilding in Spain.

Now under development is a new version of the multimedia application for doctoral students of arts. This project, funded by the Comisión interministerial de ciencia y tecnología (National Institute for Scientific and Technical Research Funding), aims to train students in specialized bilingual terminology. In addition, future projects include the design and development of a multimedia application for learning English for marine engineering and, because the European Union is promoting multilingual projects, the introduction of multimedia ESP applications in other European languages.

◈ CONTRIBUTORS

Elena López Torres is senior professor of ESP at the College of Naval Architecture and Marine Engineering, University of Cádiz, in Spain, where she has worked since 1988. She is head of the interdisciplinary research group InterDis and coauthor of *Inglés técnico naval* (3rd ed., 1998) and *Curso multimedia de inglés naval* (InterDis Research Group, n.d.)

María Dolores Perea Barberá is associate professor of ESP at the College of Naval Architecture and Marine Engineering, University of Cádiz, in Spain, where she has worked since 1994. Currently, she is writing her PhD thesis on English-Spanish maritime terminology.

CHAPTER 6

An ESP Program for International Teaching Assistants

Dean Papajohn, Jane Alsberg, Barbara Bair, and Barbara Willenborg

❖ INTRODUCTION

The number of international teaching assistants (ITAs) on U.S. college campuses has increased steadily over the years. Likewise, the testing and training of ITAs has been a growing area of concern. The ITA program at the University of Illinois at Urbana-Champaign (UIUC) addresses this concern by requiring assessment of all ITAs' language proficiency and participation in a presemester orientation. Prospective ITAs who do not pass the assessment take courses to improve their language, teaching, and cross-cultural communication skills. This chapter first describes the historical context of the ITA program at UIUC, which highlights the need for ITAs to learn the English and culture of instruction on U.S. university campuses. Next, we discuss the distinguishing programmatic and instructional features. Of particular interest is the interplay among the multiple layers of constituencies and the administrative mechanism designed to carry out ESP training for ITAs. Finally, we offer some practical ideas of possible relevance to other ESP programs, such as multilevel involvement from the organization, independent learning, individualized instruction, the use of ongoing program evaluation, realistic expectations for the program, and discipline-specific training.

❖ CONTEXT

U.S. universities have seen the number of international students increase from 12,110 in 1954–1955 (*Open Doors,* 1955, cited in Smith, Byrd, Nelson, Barrett, & Constantinides, 1992) to 211,280 in 1997–1998 (*Open Doors,* 1998). Along with this growth has come the increased use of ITAs to teach undergraduate courses. Before the early 1980s, universities typically had neither formal procedures for the screening of oral English proficiency nor provision for training ITAs in cross-cultural communication and instructional skills. This situation changed on many campuses when citizens, usually parents with children at state universities, complained to state legislators about the communication problems of ITAs (Bailey, 1984). In 1993, 17 states, including Illinois, had statewide mandates for oral English proficiency (Monoson & Thomas, 1993). For universities without state mandates, having an effective ITA program may be one way to convince citizens and legislatures that government regulation is not needed to safeguard undergraduate education.

This case study focuses on the ITA program at UIUC. UIUC began as a land-grant institution in 1868 and is now the largest public university in Illinois, with 27,855 undergraduate and 8,835 graduate students in 1999. Of the 2,734 teaching assistants (TAs) that year, 1,391 were international graduate students. The ITAs at UIUC come from China, India, Korea, and Russia as well as from other nations throughout Africa, Europe, and South America. ITAs are spread across a wide range of disciplines, such as engineering, computer science, mathematics, biology, business, education, art, and music. Within these disciplines ITAs may serve as lab assistants, discussion section leaders, lecturers, or graders with office hours. In some departments, like physics, the ITAs may teach for two semesters and then go on to be research assistants. In other departments, like English, the ITAs may teach for eight semesters or more.

In 1987 the Illinois legislature established an ITA program mandate through Senate Bill 1516, which states,

> The Board of Trustees of the University of Illinois shall establish a program to assess the oral English language proficiency of all persons providing classroom instruction to students at each campus under the jurisdiction, governance or supervision of the Board, and shall ensure that each person who is not orally proficient in the English language shall attain such proficiency prior to providing any classroom instruction to students. (Sec. 7c)

In response, UIUC's vice chancellor for academic affairs issued a policy statement regarding the appointment of nonnative-English-speaking graduate students as TAs. UIUC defines a nonnative speaker of English as someone whose mother tongue is other than English regardless of the country that person originates from or resides in. The definition includes those with dialects other than American or British English, such as English speakers from Africa and India. UIUC's ITA policy requires the assessment of oral English, the establishment of development and improvement programs, and the monitoring of oral English in actual instructional settings.

One strength of UIUC's ITA program is that the ITA policy originates from the Office of the Vice Chancellor of Academic Affairs. This upper level support ensures proper funding for the program as well as compliance from departments on campus. Without this upper-level support, some universities may end up with ITA programs that are understaffed, underutilized, or both.

◈ DESCRIPTION

The ITA program at UIUC consists of four main parts: assessment, orientation, courses, and monitoring. Assessment of ITAs is directed by the Office of Instructional Resources (OIR), which acts as a center for teaching on campus. The OIR also performs ITA development activities, such as a presemester orientation. ESL courses for ITAs are offered by the Division of English as an International Language (DEIL). The responsibility for monitoring belongs to the department in which the ITA's appointment originates. Monitoring consists of observing the new ITA while teaching and having the new ITA collect feedback from students early in the semester. A department needing assistance with an ITA can contact the OIR for additional support.

Assessment

As noted above, an international graduate student must demonstrate proficiency in oral English before becoming a TA. Each year UIUC tests more than 500 potential ITAs, using the Test of Spoken English (TSE) and its institutional counterpart, the Speaking Proficiency English Assessment Kit (SPEAK), for screening. The TSE and SPEAK are standardized tests of general oral English proficiency (Educational Testing Service [ETS], 1996). They are given via audiotape and test booklet, and responses are recorded on audiotapes, which are rated independently by two raters.

Although the TSE is available worldwide from ETS, most students opt to take the SPEAK (offered four times per year by the OIR) after arriving on campus. The holistic rating of the TSE and SPEAK is based on communicative competence theory. After faculty and students from more than 20 academic units on campus reviewed benchmark tapes, UIUC chose a cutoff score of 50 (on a scale of 20–60, indicating *communication generally effective: task performed competently*) (Papajohn, 1997). Students with borderline scores on the SPEAK may ask, with the approval of their department, to take the SPEAK Appeals test, which is coordinated by the Graduate College and consists of an oral interview and a short teaching demonstration that assesses the examinees' language skills in a teaching context. Students with scores below 50 may retake the SPEAK during another semester only after fulfilling an approved language improvement activity, such as a UIUC graduate-level ESL course, or working with a private ESL tutor for a minimum of 10 hours.

Presemester Orientation

More than 200 new ITAs representing more than 50 departments attended the August 2000 orientation. Approximately 35% were from mainland China, and 15%, from the Indian subcontinent.

A week-long presemester orientation is held each fall (August) and spring (January), the week before classes begin. The first 2 days, intended only for ITAs, address language and culture skills for teaching in U.S. university classrooms. The next 2 days, U.S. and international TAs come together to focus on teaching skills and university policy related to teaching, such as capricious grading and academic integrity. On the final day, all TAs present an 8-minute lesson that is videotaped and later analyzed one-on-one with an instructional development consultant from the OIR.

The ITA orientation begins with a large-group session entitled Drama Techniques for Communication in the American Classroom. Designed in conjunction with a professor from the theater department, the session challenges the new ITAs to think of their mind-set, their physical self, and their surroundings in addition to the content they need to teach. They cover projection techniques, ways of overcoming first-time jitters, and cultural issues such as personal space, physical contact, and body language.

After this first, large-group session, ITAs rotate through several small-group sessions. In Academic Role Plays, ITAs act out various classroom situations in pairs. They discuss cultural differences in body language, voice tone, and word choice. Voice Skills for Communication deals with word stress, phrase stress, pauses, and intonation. Classroom Communication Strategies focuses on the language appropriate for various classroom tasks (e.g., greeting or getting class started, ending class, explaining the grading system, clarifying student comments and checking understanding, inviting feedback or discussion, and collecting or assigning homework) as well as some undergraduate slang (e.g., *to ace, to bomb, to book, to cram, to blow off, to be psyched, to dis, brain cramp, Greek, rush, to nail someone*). Cross-Cultural Classroom Issues explores the role of culture in communication, addressing communication breakdowns and tactics to avoid or overcome them. In the ITA Share Session, seasoned ITAs share their experiences, field questions, and discuss strategies for dealing with a variety of classroom issues.

Interspersed among the small-group sessions are additional large-group sessions. Coherence in Teaching offers suggestions on how to effectively use words, phrases, intonation, and gestures to add coherence when communicating. In a session called International Teaching Experience, an international professor who has been recognized at UIUC for effective teaching shares cross-cultural or language challenges he or she has faced as well as strategies used to overcome the difficulties of teaching in a second language and culture. Finally, an undergraduate panel fields questions from the new ITAs and discusses topics such as fraternities and sororities, part-time jobs, extracurricular activities, and changing majors, which are often foreign concepts to an ITA.

Courses

UIUC offers two distinct ESP courses for prospective ITAs: ESL 404, English Pronunciation for ITAs, and ESL 406, Oral Communication for ITAs. Class size is limited to 12 students. Approximately two sections of each class are offered every fall (August–December), spring (January–May), and summer (June–July) semester. The classes are each considered a 3-hour course, but they are non-credit-bearing (i.e., the hours count toward full-time student standing for assistantship and visa purposes but do not contribute hours toward graduation). Therefore, courses are graded on a satisfactory/unsatisfactory basis. Each course meets 3 hours per week for 16 weeks in the fall or spring, or 6 hours per week for 8 weeks in the summer. In addition to regularly scheduled class meetings, the classes include approximately four 30-minute student-teacher conferences.

English Pronunciation for ITAs

The purpose of English Pronunciation for ITAs is "to provide rules, strategies, and contextualized practice in the stress, rhythm, and melody of English words and discourse" (Hahn & Dickerson, 1999a, p. v). To accomplish this, students focus on learning and practicing pronunciation rules for stress, rhythm, and intonation; practicing pronunciation in academic contexts; and developing skills as independent language learners.

Course content covers rules to produce proper message units, word stress, phrase stress, intonation, and rhythm, including linking and trimming. For example, when saying the word *graduate*, a student might stress the last syllable instead of the first. After learning the *left stress rule* (Hahn & Dickerson, 1999a), the student should be able to identify words with *-ate* endings and stress the correct syllable *grad-*. Another student might use incorrect stress when talking about microeconomics and macroeconomics. After learning about the placement of primary stress on parts of words to help focus the listener's attention, the student can stress *micro* and *macro*.

Students also study the patterns and placements of various intonation patterns. For instance, in giving the instruction "After you finish the first five problems, check your answers with a partner," students learn to say *problems* with rise-to-midrange intonation to indicate an incomplete thought and to pronounce *partner* with low-range intonation to mark a complete thought. Additionally, the course covers pronunciation rules for eight discourse domains relevant to classroom contexts (Hahn & Dickerson, 1999b): comparing and contrasting, lists and series, choice questions, *yes/no* questions, tag questions, information questions, narrowed questions, and repetition questions. For example, one type of comparing and contrasting is *X, not Y,* where *X* and *Y* carry the primary stress. Therefore, an ITA who wants to say, "This is your responsibility, not mine," should place primary stress on *your* and *mine.*

After rules are presented in class, students apply them to dialogues and conversations. During class the teacher rotates among pairs of students to give targeted feedback. The teacher models correct forms and offers suggestions on changes students should make. If the class is working on the linking of sounds and the instructor hears a student say, "They tried developing an alternative solution" with the words *tried* and *developing* separated by a vowel sound, the instructor uses this as a teaching moment. The instructor can identify consonant-to-consonant linking between the *d* of *tried* and the *d* of *developing* and give advice about not overreleasing air with the *d* in order to link the words.

The ESP context of teaching in a U.S. university is an integral part of the course. Early in the semester, each student creates a general academic terms list—composed of words such as *analysis, approximation, characteristic, development, hypothesis, phenomenon, qualitative, quantitative, subsequent, technology,* and *variable*—and a specific academic terms list. Students in electrical engineering may choose words like *electron, Kirchhoff's Law,* and *microprocessor,* whereas students in biology might list words like *erythrocyte, plasma membrane,* and *photosynthesis.* As students learn the four word-stress rules (i.e., the key, V/VC, left, and prefix stress rules), they apply them to their own lists of words.

All discourse-level practice exercises are contextualized, for example, in situations involving instructors talking with students or with other instructors, using questions in the classroom, or presenting lessons. A practice dialogue might take the form of a student-teacher interaction in office hours:

Student:	I just can't solve this problem.
TA:	Why don't you tell me what you've tried so far?
Student:	I've written the force equations and tried to plug in the numbers, but I still don't get anything.
TA:	Have you tried drawing a free body diagram?
Student:	No.
TA:	Why don't you try that?

Using dialogues like this one, students practice primary stress, linking, question intonation, or the rhythm of contractions.

Students also prepare and deliver field-specific talks. Topics may include defining a term, describing a process, or summarizing. The first step is to write the text of the talk. Next, the message units are marked with a slash, and the phrase stress is marked by a large circle or capital letters. This excerpt is from a script written by a student in electrical engineering:

> hello CLASS / today I would like to introduce a basic TERM / digital SIGnal process / FIRST of all / what IS a signal / SIMply speaking / it is a series of changing VAlues / along a certain INdex / for exAMPLE / a signal can be the temperature outSIDE / the temperature changes everyDAY / ANY kind of changing value / can be LOOKed at/ as a SIGnal / . . . .

To develop self-monitoring skills, students rehearse their talk at home three times before making an audio recording. Students focus on pronunciation features such as message units, intonation, linking, primary stress, word stress, compound nouns, vowel and consonant sounds, pacing, and volume (Hahn & Dickerson, 1999b). For example, a student might notice the compound noun *signal process* and realize the need to stress *signal* instead of *process*, or notice a pause between *would* and *like,* which means the student is not forming grammatical message units. Or a student might notice that the words *series, changing,* and *values* are all receiving equal stress instead of *values* receiving a bigger pitch move to mark the phrase stress. The presentation is recorded a second time in class. Finally, using self-, peer, and instructor feedback, students make a third recording at home. This process of rehearsal and analysis is known as *covert rehearsal* (Hahn & Dickerson, 1999a). Self-monitoring and self-correction skills equip students to keep on improving after the class is over. Preliminary survey data suggest that students who continued to employ covert rehearsal strategies after the end of the course noticed improvement in their English pronunciation (Alsberg, in press).

Oral Communication for ITAs

The purpose of Oral Communication for ITAs is to prepare current and potential ITAs in teaching, language, and cross-cultural skills for the U.S. university classroom. The ESP nature of the course is reflected in the syllabus, which is organized around videotaped teaching tasks, such as introducing oneself, explaining a field-specific visual, fielding questions, defining a field-specific term, role-playing office hours, and leading a discussion. For each videotaping, students may evaluate a model lesson, create and practice a sample lesson, teach the lesson, analyze their own videotape, and discuss the videotaped lesson in a conference with the instructor. Feedback is considered crucial to improvement, so the course incorporates multiple sources of feedback, including the teacher, undergraduate students, peers, and the student's self-evaluation.

In Oral Communication for ITAs, students work on each of the components of communicative competence. The course addresses functional and strategic competencies by assisting ITAs in accomplishing the teaching tasks described above and through role plays, such as managing unusual classroom situations. One such situation might be "A student asks you a question, but you do not know the answer. What should you do? What should you say?"

Linguistic competence is addressed both in and out of class (in conferences or with homework assignments) in the areas of pronunciation, grammar, and vocabulary. Early in the semester, instructors identify pronunciation problems with a diagnostic assessment consisting of a paragraph reading and open-ended questions. Topics that are difficult for most of the ITAs, such as phrase stress and intonation of message units, are covered in class and practiced on tape at home. Instructors help students with individual pronunciation problems in conferences and through individualized help based on their own presentations. Because questioning skills are an important aspect of ESP training for ITAs, the prospective ITAs study the grammar and intonation of questions. The formation of embedded questions is especially difficult for students. For example, the teacher may ask the ITAs to change the form of a question from "What does *acceleration* mean?" to "Can anyone tell me what *acceleration* means?" As follow-up, students do written exercises, make audiotapes, and analyze their videotaped lessons for correct use of embedded questions. Additionally, the instructors note recurring grammatical errors in the videotaped lessons and share them with students during conferences. When appropriate, instructors assign individual grammar exercises. The class also covers vocabulary and expressions that are relevant to the context of the classroom (Smith, Meyers, & Burkhalter, 1992). When responding to incorrect answers, an ITA can say, "Do the rest of you agree? Is that right?" or when giving hints to students, an ITA could say, "If we . . . , what will happen?" or "Why did you . . . and not . . . ?" The prospective ITAs practice language for student-teacher interaction through role plays.

Instructors address discourse competence by assigning exercises that focus on organizational cues (J. Smith et al., 1992) and teaching ITAs to organize their lessons cohesively The functions of transition words are presented: numeric transitions such as *first* and *second*; sequential transitions such as *after* and *then*; additive transitions such as *and* and *also*; and summary transitions such as *so* and *finally*. To teach sociolinguistic competence, the instructor might have the class look at direct and indirect expressions and discuss the differences between "Why didn't you use the chart in the back of the book?" and "You might want to use the chart in the back of the book to solve this problem."

Because many students in the course are concerned about taking the SPEAK, throughout the semester the program holds several SPEAK workshops. Using exercises in *Toward Speaking Excellence* (Papajohn, 1998, pp. 48–51), students work on their fluency and organizational skills. For example, students learn the importance of cohesive devices in showing logical connections between ideas and practice using these devices. A student might say, "The explanatory power of the regression model adopted in Table 14 is 22.9%, *and it is reasonably high*" instead of *which is reasonably high,* or "There has been a tremendous volume of research which shows that the information content of earnings *and rapid stock price movements*" instead of *is related to rapid stock price movements.* Students can become aware of and practice the accurate use of cohesive devices by narrating picture stories or presenting schedules.

In addition to focusing on language skills, the course strives to sensitize ITAs to the cross-cultural issues of teaching in the U.S. university classroom. Learners may compare the qualities of a good teacher in their own culture to those of a good teacher in U.S. culture, or interview U.S. students to learn what they think the characteristics of a good teacher are (J. Smith et al., 1992). The interactive nature of the U.S. classroom is also highlighted. Many ITAs come from cultures in which students rarely interrupt the teacher for questions during class. An openness to students' questions and the fostering of active participation in class is a cross-cultural attitude that can affect an ITA's success in the U.S. university. Because international graduate students often have little contact with U.S. undergraduate students, instructors occasionally invite U.S. undergraduate student volunteers to ask questions and give written feedback during ITA teaching presentations. Undergraduate students and ITAs also discuss student life on campus.

A more recent development at UIUC is to offer discipline-specific training for international gradute students who will work in departments that utilize significant numbers of ITAs. In spring 1999, DEIL piloted an oral communications course for the Department of Theoretical and Applied Mechanics (TAM). Language and teaching skills were supplemented with discipline-specific content from lab manuals and engineering texts. Students used technical definitions for terms such as *flexural*, *compressive*, and *tensile strength* to practice proper pausing and phrase stress. To practice the language and technique of describing graphs, instructors brought to class lab materials that the ITAs would eventually teach. These materials were not straightforward bar graphs but technical graphs with titles like "Relation Between Creep Exponent M and Possible Deformation Mechanisms in Metal," "Fracture-Toughness Load-Deflection Curve," and "Calibration Chart for Concrete Made With Crushed Limestone and Natural Sand Aggregates." ITAs practiced word stress and segmental sounds for field-specific terms according to their interests and the classes they would soon teach. For example, ITAs who would teach fluids labs practiced pronouncing vocabulary like *transducer, streamline,* and *viscosity*; ITAs who would teach statics courses practiced *brittle, fracture,* and *coefficient of friction*; ITAs who would teach materials courses practiced *porosity, air-entrained,* and *shear-slump.* For additional language practice outside the classroom, learners paired up with U.S. undergraduate conversation partners from TAM, which gave the ITAs a chance to explain how to solve engineering problems, define engineering terms in a way that undergraduates could understand, and learn some undergraduate slang, such as *boss, evil, nuke,* and *razz.*

To analyze relevant communication and teaching skills, students watched videotapes of U.S. TAs and ITAs teaching TAM courses. Below is a short transcript segment from a videotape of a U.S. TA teaching a TAM class.

> you're supposed to use uh you're we're we're studying these forces instead of the bearing forces—at some point they're actually its kind of equivalent 'cause these guys are actually what what what this feels—if this is a perfect weld so you don't have to worry about this breaking then all the forces here are actually felt at the ends and its holding the same wobbling—yeah

The ITAs discover a number of points from this short transcript. Although the TA is fluent in English, the ITAs noticed that the delivery was not very smooth or efficient. The TA used a great deal of unnecessary language, and there were several distracting hesitations and repetitions. Yet these did not prevent the TA from communicating clearly; the ITAs could observe this from the video. Additionally, the U.S. TA used contractions and reductions like *you're* and *'cause* to maintain the rhythm of his language. The effect of using inclusive language like *we're* instead of *you're* was also discussed. Informal expressions like *kind of, these guys,* and *wobbling* were examined in context. This type of video and transcript greatly helped the ITAs because they could observe and learn from the language used by teachers in the classes they were soon to teach.

◈ DISTINGUISHING FEATURES

The distinguishing features of the ITA program at UIUC can be divided into two categories: program features and instructional features. Program features relate to mandated aspects of the program and the various constituencies involved in fulfilling the mandate. The instructional features relate to the teaching methods and materials, program evaluation, and the integration and scope of the various aspects of the program.

Program Features

Meeting Legal Mandates

The state law requiring all instructors at UIUC to be fluent in English creates a legal motivation for UIUC's ITA program. Other professions, such as medicine, nursing, physical therapy, and engineering, may have similar mandates. The existence of the law distinguishes ESP programs for these professionals from ESP programs for landscapers, factory workers, and hotel staff, who are encouraged to improve their language in order to improve job effectiveness but are not fulfilling legal mandates on language proficiency.

At UIUC the state mandate prompted the formation of an upper-level administrative ITA policy that resulted in standardized assessment utilizing the TSE and SPEAK. Although the TSE and SPEAK provide a consistent, efficient way to assess large numbers of prospective ITAs, they do not necessarily predict if someone will be a good TA. Other factors such as knowledge, flexibility, cross-cultural adaptation, availability, and commitment can affect the overall effectiveness of a TA. For this reason, the ESP courses and ITA orientation include teaching and cross-cultural skills.

Responding to Multiple Constituencies

Beyond meeting a legal mandate, UIUC's ITA program endeavors to take into account multiple constituencies, including campus administration, departments, professors, graduate students, undergraduate students (and their parents), and the state. Although some ESP programs may operate in isolation of the overall organization, UIUC's ITA program balances the needs of each of these constituencies and works to integrate them in the execution of the program.

The state mandate and the ITA policy statement from the Office of the Vice Chancellor of Academic Affairs provide the upper-level support needed to ensure compliance from departments and ITAs. The decision on a SPEAK cutoff score involved faculty and students from departments campuswide. ESL courses are funded by the Office of the Vice Chancellor for Academic Affairs, so prospective ITAs who do not pass the SPEAK have the opportunity to improve their language skills. Faculty from the international student's department, as well as representatives from the Graduate College, DEIL, the OIR, and undergraduate students, are involved in the SPEAK Appeals test. The OIR, DEIL, various faculty members, ITA mentors, and undergraduate students all have a part in the ITA orientation. This multilevel involvement keeps the ITA program visible to the various constituencies.

Instructional Features

ITA-Specific Content, Materials, and Methods

ITA training is teacher education with a special focus on language and culture. Thus, ITA trainers view language teaching from the perspective of how it fits into the context of the U.S. university classroom. Language is just one tool for accomplishing teaching tasks. Teaching techniques and cross-cultural skills are additional tools that can compensate for linguistic deficiencies and enhance communication.

The content of the ESP courses and orientation is specific to TAs. Prospective TAs learn functional language associated with the tasks teachers face, such as defining terms, explaining grades, and fielding questions. Vocabulary common to academia, to specific disciplines, and to undergraduate students is also incorporated into the program. Prospective TAs learn language with which to actively engage their undergraduate students by leading discussions, facilitating problem solving, and negotiating meaning. Videotaping and audiotaping technology are used extensively for teaching.

The ESP courses rely on materials and methods developed at UIUC. Publication of *Speechcraft: Discourse Pronunciation for Advanced Learners* (Hahn & Dickerson, 1999a), *Speechcraft: Workbook for International TA Discourse* (Hahn & Dickerson, 1999b) and *Toward Speaking Excellence* (Papajohn, 1998) makes this information available to other programs.

A basic premise underlying the ESP courses is that a large portion of language improvement occurs outside the classroom. Therefore, the ESP classes include instruction in methods for self-evaluation that learners can apply even after the classes end. Students learn covert rehearsal and are asked to evaluate their own language and the language of their peers. The instructor gives language feedback to learners not only to help them identify particular problem areas, but also to illustrate how students can apply the same techniques in self-evaluation. For example, after listening to a student's audiotape, an instructor may comment on accurate compound noun stress and inaccuracies in contrasting stress. This provides a model for the student to use when evaluating his or her own audiotape in the future or when evaluating the audiotape of another ITA. With concrete feedback, students will more likely focus and comment on specific aspects of pronunciation rather than generalize and merely say the pronunciation is "pretty good" or "needs some work."

Ongoing Program Evaluation

Program evaluation is an ongoing process in UIUC's ITA program. Each semester, instructors of the courses collect written feedback from students halfway through the semester. This feedback can lead to adjustments or serve as an impetus to explain the rationale behind certain learning activities. For example, instructors may need to address the misconception that, as nonnative speakers of English, international students are not qualified to analyze and give feedback regarding their own or their classmates' oral English. Students may be unfamiliar with or may misunderstand the concept behind communicative language activities as well. Explaining the purpose or focus of specific activities may prove helpful in getting students to participate in activities enthusiastically.

The program continually fine-tunes the course delivery, materials, scheduling, assignments, and activities based on analyses of end-of-semester student evaluations.

The ITA orientation also collects feedback from ITAs and staff to keep the program from becoming stagnant. For example, over the past decade, the language skills of new ITAs have improved. Prompted by feedback, the type of language training offered at the orientation has been adjusted. Feedback has also led to the elimination of areas of overlap between various sessions of the ITA orientation as well as the revision or elimination of unsuccessful activities.

Comprehensive Support for ITAs

A final distinguishing feature is the comprehensive nature of UIUC's ITA program. International students register for either an ESP course (one of two levels) or the ITA orientation based on their SPEAK score. One course offers help in basic pronunciation while another assists more advanced learners in developing confidence and improving their skills as instructors. Individual departments and the OIR provide follow-up for ITAs once they have begun teaching. This follow-up may include live or videotaped observation, one-on-one or group consultation, or instructional workshops. ITAs and potential ITAs are supported at various points along the course of their professional development due to the broad scope of the ITA program.

◈ PRACTICAL IDEAS

The successful aspects of UIUC's ITA program as well as some unfulfilled needs of ITAs suggest some practical ideas for ESP work.

Encourage Multilevel Involvement

Multilevel involvement means organizationwide involvement in various aspects of the program, specifically with upper-level support. Without this support, an ESP program is vulnerable to budget cuts; without organizationwide involvement, compliance and cooperation with the program may become problematic. When an ESP program is highly visible, potential candidates for the program may be more inclined to participate and to continue with the training. Furthermore, when people within an organization run into problems that the ESP staff can address, they immediately know where to go for ESP assistance.

Equip Students to Be Independent Learners

No ESP program can teach learners everything they need to know. Because a great deal of language learning occurs outside the classroom, equipping students to be independent learners is important. Creating opportunities for conversation partners or other language practice experience outside the classroom can enhance language learning.

Individualize Instruction

Limiting class sizes, performing diagnostic assessments, setting individualized language learning goals, and holding one-on-one conferences benefit the learners and help them see the personal value in an ESP program. Videotaping and audiotaping followed by a discussion and analysis are useful tools in individualizing instruction.

Evaluate the Program

Ongoing evaluation is necessary to keep a program focused and up-to-date. If an ESP program does not meet the current needs of its learners, motivation and participation dwindle. Furthermore, program evaluation is essential in demonstrating the benefits of the ESP program to decision makers.

Program evaluation can also help identify why some learners benefit from a given ESP program whereas others do not. Recently, preliminary feedback from students and instructors has indicated that some UIUC students need to improve in the areas of grammar and fluency before they can benefit from the existing ITA program.

Have Realistic Expectations

It is important to identify the needs an ESP program can and cannot address, and to communicate this information to the ESP program sponsor. The ability to hold realistic expectations comes from needs assessments and experience. For example, at UIUC, ITAs may share the common context of teaching but work in disciplines as diverse as music and physics and be assigned very different duties, such as facilitating discussion sessions, leading labs, or holding office hours. Because a single program cannot meet all of the ESP needs in these contexts, the university must be realistic about what can be accomplished.

Another area requiring realistic expectations concerns cutoff scores set by universities on tests of English proficiency. If the cutoff score is unnecessarily high, then competent international graduate students may be prohibited from teaching. If the cutoff score is unreasonably low, then some international graduate students who are not linguistically ready for the challenge of teaching may become TAs.

If English assessment is mandated, program staff should thoroughly consider the ramifications of cutoff scores and the opportunities for improvement available to potential TAs who do not reach them. It may be unrealistic to expect some international graduate students to improve their oral English without providing the structure for them to do so. As mentioned above, one strength of UIUC's ITA program is its breadth—from screening, to service courses, to presemester and on-the-job training.

Employ Discipline-Specific Training

Grouping learners by academic discipline has a number of benefits. For example, it allows the integration of discipline-specific materials ranging from technical terms to graphs and lab reports into the program. Such authentic materials provide realistic practice and motivation. Another benefit is that instructors can use discourse analysis from the discipline to show examples of effective and ineffective communication. If the ESP instructors happen to have some background in the given discipline, they can better discern whether learners can or cannot communicate discipline-specific content well and have a better understanding of the generally accepted methods of communication in that discipline.

Some disadvantages to grouping learners by academic discipline exist as well. One is the potential lack of freshness and variety in the classroom if everyone shares the same knowledge and interests. Another potential drawback is that shared background knowledge can bias the peer feedback of an in-class presentation. Peers from the same field may not give feedback that emulates that of undergraduate students because they already know much of the content in the presentation. In either case, however, ESP instructors need to assess their learners' needs and group students in ways that will help them the most.

◈ CONCLUSION

Although each ESP program is situated in a unique context, similarities exist among programs. Presentation skills taught to ITAs are similar to those taught in the ESP training of business executives, and questioning instruction for ITAs can be compared to training that medical personnel receive for question-and-answer exchanges with patients. Multilevel involvement, independent learning, individualized instruction, program evaluation, realistic expectations, and discipline-specific training are relevant to a wide range of ESP programs.

In the future, the ITA program at UIUC needs to investigate further opportunities for improvement. ITA courses within specific disciplines, such as the ESP course for the Department of Theoretical and Applied Mechanics, offer much promise. Related to discipline-specific instruction is the need for more discourse analysis. Research in discourse analysis can inform instruction in language functions, vocabulary, syntax, repair strategies, and interaction patterns. Field-specific testing is of future interest as well. Many universities already utilize teaching performance tests on a larger scale than the SPEAK Appeals test administered at UIUC. Another area for investigation is the development of courses to round out the existing offerings for ITAs, such as courses in grammar and vocabulary for improved fluency. Although much has been accomplished, much remains to be done.

◈ CONTRIBUTORS

Dean Papajohn is a teaching associate in the Division of English as an International Language (DEIL) and an educational specialist in the Office of Instructional Resources (OIR) at the University of Illinois at Urbana-Champaign (UIUC), in the United States. He has worked with ITAs for more than 10 years.

Jane Alsberg is an educational specialist in the OIR at UIUC. She taught French and Spanish as foreign languages to elementary through high school students for more than 8 years before coming to UIUC, where she has been working with ITAs for more than 5 years.

Barbara Bair currently teaches ESL at Urbana Adult Education, Urbana, Illinois, in the United States. From 1993 to 2001, she taught ITAs and MATESL candidates in the DEIL at UIUC. She has taught EFL in Zambia, Nigeria, and Taiwan.

Barbara Willenborg is the ITA coordinator at the University of Pennsylvania, in the United States. She has been working with ITAs for more than 4 years.

PART 2

ESP for Language Learners in the Workplace

CHAPTER 7

An ESP Program for International Medical Graduates in Residency

Susan Eggly

◈ INTRODUCTION

Medical literature reflects an increasing awareness of the relationship among physician-patient communication and satisfaction, compliance, medical outcome, and malpractice suits against physicians (Cole & Bird, 2000; Silverman, Kurtz, & Draper, 1999). Awareness of the key role of physician-patient communication has led to an increase in communication skills training programs in medical schools and residency training programs throughout the United States. An important issue to consider in the development and implementation of these programs is that many physicians in training in the United States were born and raised elsewhere and attended medical school in their native countries. In fact, 26% of all residents in U.S. residency training programs in 1997 attended medical school outside the United States and Canada (Dunn, Miller, & Richter, 1998). Although this group of physicians, known as international medical graduates (IMGs), includes U.S.- or Canadian-born physicians who have gone elsewhere for medical school, the majority came to live in the United States for the first time after graduating. The majority of these eventually established their career in the United States, making up approximately 23% of the physicians practicing there (Inglehart, 1996).

Before applying for a U.S. residency training program, IMGs must be certified by the Educational Commission on Foreign Medical Graduates (ECFMG). Until recently, the certification process included an objective English language proficiency exam similar to that required by most universities. In 1998, however, an exam known as the Clinical Skills Assessment was instituted as a part of the requirements. Through the assistance of actors playing the role of standardized patients, this rigorous performance exam tests clinical as well as interpersonal and language skills. Because the test became a requirement recently, however, only IMGs who have applied for ECFMG certification since 1998 have taken the exam; consequently, most IMGs currently in residency training programs or in practice have not demonstrated their ability in interpersonal and language skills.

The ESP community has a rare opportunity to contribute to the improvement of physician-patient communication through instruction in language and culture in U.S. medical settings. This chapter describes a program that offers instruction in language, culture, and physician-patient communication skills to IMGs in a large, general internal medicine residency program in an urban medical center in Detroit,

Michigan, in the United States. Components of the course include cultural orientation, interviewing skills instruction, and private language tutorials.

◈ CONTEXT

On July 1 of every year, IMGs and their U.S. counterparts join residency training programs all over the United States. The Wayne State University general internal medicine residency training program is typical in that residents rotate through a variety of clinical settings each month for 36 months, caring for hospitalized and clinic patients who suffer from illnesses such as sickle cell disease, HIV, diabetes, hypertension and congestive heart failure. Many patients have medical problems related to unhealthy lifestyles, such as homelessness, substance abuse, or social isolation. During some rotations, residents work as many as 100 hours per week, including 30-hour shifts known as *call* every fourth day. A few rotations, however, are more relaxed, and residents have evenings and weekends off. The first year, known as the internship, is the most intense in terms of workload. During the second and third years, residents take on more leadership and teaching responsibilities.

Before entering a residency training program, residents must have graduated from an accredited medical school. Some IMGs practice medicine in their home countries for many years before deciding to come to the United States to further their training. Of the approximately 100 residents in the Wayne State training program during any given year, on average fewer than 5 were born and raised in the United States. In the 1998–1999 academic year, for example, more than 50% came from India; nearly all the others came from China, Pakistan, Iran, Iraq, Korea, the Dominican Republic, Syria, and Jordan.

Medical residents from India have typically received all of their education in English and are generally proficient in nearly every aspect of the language except, in certain cases, American English pronunciation. There are several problems, however, with the spoken English of many of the Indian physicians. First, they have learned British rather than American English; second, they have used English in very formal, professional settings and are less familiar with informal registers or the medical language of nonprofessionals; and, third, they are generally unfamiliar with the English used by their patients, most of whom are African American. Many problems result; for example, in referring to their medical problems, patients frequently use slang that their physicians find confusing. At the same time, the formal, British English used by Indian IMGs sometimes intimidates or confuses the patients.

Residents from countries other than India have typically been educated in their native language and have little experience using English outside of their English class. Although they may be familiar with medical terminology and have excellent reading skills gained from reading English language medical journals, they may not be accustomed to using English in a medical or social context, such as presenting a case to a faculty physician, explaining a procedure to a patient, listening to a lecture in English, or even ordering a sandwich in the hospital cafeteria. Additionally, they are generally unfamiliar with the holidays, celebrations, and traditions of their patients and colleagues; in fact, some have cultural norms that are quite contrary to aspects of lifestyle in the United States, such as in the areas of alcohol consumption or extramarital sexual relationships.

◈ DESCRIPTION

The curriculum I created to address the needs of IMGs in the areas of language, culture, and communication skills in the Wayne State University residency training program has three components: cultural orientation, interviewing skills, and individual language tutoring. I developed the curriculum over the years 1990–2000 after a brief needs analysis by several members of the Division of General Internal Medicine, including several physicians and me as the ESP/communication skills instructor. Results of the needs analysis indicated several areas in which IMGs reported feeling uncomfortable, among them adjustment to the hospital system (e.g., use of the computer system, roles of and relationships with faculty and nurses, use of acronyms and slang, structure of the office visit); adjustment to daily life (e.g., driving, shopping, eating, coping with homesickness); and the sociocultural background of patients (e.g., family structures, alcohol and drug use, financial issues). The curriculum has evolved over the years in response to the changing needs of the residents. Each component is organized and supported by the Medical Education Office of the Department of Medicine and delivered by a core group of faculty in the Division of General Internal Medicine. Following is a detailed description of each component of the curriculum.

Cultural Orientation

Cultural orientation occurs during the first 2 months of the first year of the residency. Because the residency program begins officially on July 1, most residency programs, including Wayne State's, provide intensive orientation during the last few days of June. Most of the orientation is devoted to facilitating the new residents' learning about the hospital system: learning to access patient records and other information from the computer system; receiving lab coats, beepers, and badges; meeting faculty; filling out forms; getting to know each other; and a variety of other tasks. Three components deal directly with cultural adjustment during the orientation: a discussion on multiculturalism, a bus tour of the surrounding area, and a series of noontime lectures on a variety of topics.

Discussion of Multiculturalism

A discussion of cultural values in a medical setting addresses mainstream U.S. values, such as individualism, egalitarianism, and personal control, and ways in which these values influence medical issues, such as the disclosure of diagnostic information, physician-patient relationships, and willingness to adhere to medical treatments. In addition, the discussion serves as an opportunity for residents to examine cases in which dissimilarities in ethnic background have resulted in differing explanations of illness and disease and have led to conflict in a medical setting. In one case, a Mexican couple refuses to allow their infant son to be disconnected from the tubes that are sustaining his life. Residents are encouraged to consider religious and cultural issues that might give insight into reasons for the conflict and ways to handle this sensitive communication challenge. (The theoretical bases and the majority of the cases for this discussion are derived from Galanti, 1997.)

Bus Tour

The second component of cultural orientation during the first week is a bus tour of the area surrounding the hospital complex. Because many residents from outside Detroit bring with them fears and negative attitudes toward the patient population and lifestyle in the Detroit area, the bus tour counteracts these attitudes by providing a realistic portrait of the area and its inhabitants. Residents see a city of contrasts as the bus travels through a great variety of neighborhoods, some with elegant, well-kept homes and some with homes that have been abandoned and left to burn; residents see prostitutes and drug addicts on the streets as well as healthy, happy families and children walking to school. The bus passes by some of the homeless shelters and drug rehabilitation centers to which residents will refer their future patients; in addition, residents see restaurants, museums, and parks where they can take their friends and family. Residents respond with great enthusiasm to the bus tour, which ends at a riverfront park for a picnic with faculty and current residents. They comment that as a result of this experience, they are less fearful of the city and more comfortable talking with patients about their personal lives.

Lecture Series

Throughout July and August, interns continue the orientation process through a series of lectures held once a week at noon. I lead some of the lectures, and faculty physicians lead others. Lunch is served, and residents learn very quickly to overcome their cultural bias against eating in public or during meetings; in fact, they frequently refuse to attend noon lectures if lunch is not served. Following are some of the topics addressed during these lectures.

- Health care insurance plans are especially confusing to individuals who come from countries in which health care is free. This lecture covers large, public insurance systems such as Medicaid and Medicare in detail and explains some of the smaller, private systems.

- The structure of the office visit is quite different in the United States than it is in other countries. Each country has its own system; residents report that in many countries, patients wait in a long line to see the doctor rather than calling for an appointment, so waiting for the doctor is an expected part of the system. All the patients are seen in one large room; there is little in the way of charting or documentation; the visit focuses more on the present illness than on the family or social history; patients receive a very cursory physical exam and are sent on their way with either a prescription or orders to be admitted to the hospital. Consequently, residents who were used to seeing 20 patients in a morning in their own countries are overwhelmed by having to see 2 or 3 patients in a morning session in their clinic in the United States because of the detailed history and documentation requirements. This lecture highlights these differences and is accompanied by videotapes and role plays for practice.

- Prescription writing and charting can be challenging to residents who have never practiced in the United States. In many countries, prescriptions are used only for controlled substances, such as narcotics; in addition, residents may not be familiar with trade names for drugs, referring to them instead by their pharmaceutical names (e.g., acetaminophen for Tylenol). As noted above, in many countries charts do not exist, so residents need specific instructions on how to document their patient interactions, both to communicate with other physicians and to avoid legal problems. This lecture is long and tedious, but necessary.

- Alcohol and substance abuse occurs in most but not all countries of the world. Some residents have never tasted alcohol or attended a gathering where alcohol is served and have strong biases against anyone who uses alcohol. They are often unfamiliar with the notion that alcohol can be used in moderation; they are also unfamiliar with slang terms for drugs and alcohol and have difficulty assessing use versus abuse. The lecture devoted to this topic includes a video of a liquor store as well as examples (not samples) of a 40-ounce bottle of beer, a fifth of vodka, a pint of Mad Dog 20-20, and other items commonly used by patients who may or may not have related medical problems.

- Variations in language used by patients, in particular African American English, and slang can be confusing to residents who speak English as a second language. At the same time, the medical jargon used by physicians can intimidate or confuse patients. A lecture addressing these topics uses many examples of both and gives residents the opportunity to translate back and forth between the two language styles. For example, patients refer to *the runs* (diarrhea), *falling out* (fainting) or having a *significant other;* doctors refer to *benign procedures, local anesthetic,* or *peripheral neuropathy.* This lecture also includes an explanation of African American English, encouraging residents to acknowledge its status as a variety of English with regular patterns of grammar and pronunciation rather than as bad or broken English spoken by the uneducated.

Other topics include relating to nurses and other staff, taking sexual histories and performing breast and pelvic exams, and understanding community resources. Each year, the topics are adjusted to meet the needs of the incoming class.

Interviewing Skills Course

The second major component of the communication skills program is a course in interviewing skills. I teach this course to small groups, with frequent, short sessions throughout the first year of the residency, totaling 12–15 hours of instruction. Methods include readings, lectures, discussion, videos, role play, and the opportunity to be videotaped performing a simulated interview with an actor, followed by a one-on-one review with me. I adapted the course curriculum from the three-function model of the medical interview, which defines the main functions of the interview as (a) building the relationship, (b) assessing the patient's problems, and (c) managing the patient's problems (Cole & Bird, 2000). Following is a description of each of the six main components of the course.

The Physician-Patient Relationship

The physician-patient relationship, like all social relationships, is heavily laden with cultural norms and values. In a society that places a great value on hierarchy, for example, physicians may be treated with great respect; their knowledge or advice is not questioned, especially in public. On the other hand, in the United States the social and professional hierarchy is not as closely adhered to as it is in many other societies. Patients in most cases view the doctor-patient relationship as a partnership—sometimes even a business transaction—and believe that they have equal rights to information and decision making. Therefore, the interviewing skills class covers topics, such as culture and values, that influence the doctor-patient relationship, cross-cultural images of doctor and patient, and the current medicolegal environment.

Function 1: Building the Relationship

Although frequently omitted from the medical school curriculum, a strong rapport between the doctor and the patient is the foundation on which the interview is based. Most physicians admit that proper diagnosis and treatment is quite difficult when there is a lack of trust. The key to building trust and rapport is not only to feel empathy but also to show it. Therefore, residents need to understand and respond, at least superficially, to the emotions their patients express during the medical encounter.

During this segment of the interviewing skills course, residents discuss the way they express emotions as contrasted with the ways in which their patients may express emotions. Some residents may be very uncomfortable, for example, with raised voices or swearing as an expression of anger or withdrawal as an expression of sadness. They learn and practice ways to show empathy, such as reflecting emotions (e.g., "You seem very upset by your illness.") or expressing personal support (e.g., "I'm here to help you in any way I can.").

Function 2: Assessing the Patient's Problems

The nature of biomedical diagnosis requires the physician to gather a great deal of detailed information from a variety of sources, but primarily from the patient. Furthermore, in internal medicine, the history usually includes not only medical but also social and health maintenance information. In many medical settings around the world, because physicians see many more patients and have access to fewer resources for diagnosis and treatment than in the United States, the medical interview is very brief, focusing on the current problem. This section of the medical interviewing course includes topics such as greeting the patient, using attentive nonverbal behavior, organizing and setting priorities for the interview, balancing open- and closed-ended questions, and listening actively.

Function 3: Managing the Patient's Problems

Physicians frequently have difficulty explaining medical conditions in lay terms, especially if patients in their home countries are unaccustomed to requesting detailed explanations. This difficulty can be exacerbated by lack of familiarity with lay medical terms or with social and cultural issues that may interfere with the patients' ability to adhere to a treatment plan. For example, patients who lack medical insurance or transportation may be offended by a physician's suggestion of a variety of diagnostic tests requiring frequent visits to the hospital, especially if they do not have a clear understanding of the reason for the tests.

In the interviewing skills class, residents practice giving clear explanations in lay terms and checking to see if their patients understand. In addition, they explore further the concept of the doctor-patient partnership in order to learn to negotiate rather than dictate treatment, increasing the potential for adherence to the treatment plan. Role plays give them the opportunity to practice explaining a variety of diagnoses and negotiating treatment plans.

Physicians naturally feel uncomfortable providing news of a diagnosis such as a terminal or stigmatized disease. In many countries and ethnic groups, this information is communicated to the patient's family; in fact, giving bad news to a patient may be considered unethical because it is perceived to hasten the illness process. In the interviewing skills class, residents discuss the cultural and ethical implications of delivering bad news as well as the appropriate language skills for such a highly emotional interaction. Some of these skills include choosing an appropriate time and place to talk with the patient, giving a basic diagnosis using nontechnical language, eliciting and responding to patients' emotions regarding their diagnosis, listening actively, offering hope, and providing only necessary details rather than overloading the patient with extraneous technical information (Eggly et al., 1997).

The Social History

The purpose of the social history in the medical interview is to determine social influences, such as occupation, smoking, use of alcohol or other substances, marital status, support systems, and sexual activity, on patients' medical conditions. IMGs and other physicians who do not share the social background of their patients can easily offend patients or miss important information because of their personal biases against or lack of awareness of their patients' lifestyles. For example, many residents have reported that in their countries, a married person is assumed to be sexually monogamous, have children within the marriage, and participate in a mutually supportive relationship. Deviations from this social rule are considered shameful and are not publicly acknowledged. Physicians therefore consider probing into sexual activity or number of children to be extremely rude once patients have stated their marital status. In the United States, however, it is appropriate for physicians to explore issues such as sexuality, children, or abuse regardless of patients' marital status.

Videotape and Review

Following the interviewing skills course, residents are videotaped conducting an interview with a professional actress who portrays a patient. I review the tape with residents in order to give individual feedback on interviewing style. Residents are then videotaped in the outpatient clinic with real patients at least twice a year throughout their 3 years of residency training. This allows residents to continue to develop doctor-patient communication skills and to ask questions about their interactions in a one-to-one setting.

Individual Language Tutoring

The third major component of the communication skills curriculum, individual language tutoring, is the one in which the ESP professional is likely to feel the most competent but possibly the most frustrated. Individual residents seek the help of an ESP professional for two reasons: Either they must attend private sessions because of poor evaluations by faculty, or they request individual tutoring because they perceive a need. In the first case, the resident may have additional problems, such as weak medical knowledge, poor interpersonal skills, or personal problems. Residents who seek help on their own, on the other hand, are usually highly motivated. In both cases, however, I conduct a complete language needs analysis including an oral proficiency interview and a discussion about perceived deficiencies in communication.

The following example illustrates the issues and suggests ways to address them through private language tutorials with IMGs. During her first rotation in July, Dr. R, a Dominican resident, was reported to the program director and to me as having language problems that were interfering with her ability to perform her responsibilities as a member of a medical team in the hospital. Her supervisors reported that they could not assess her medical knowledge because her language skills were so weak and that, as a result, she was very insecure in her work. She was more and more nervous each day; she was becoming depressed, working nearly 100 hours per week and trying desperately to keep up with her reading, her sleep, and her family. Therefore, I designed a short-term ESP program to meet her immediate needs during the current rotation and a longer term program to take place during rotations with less intensive schedules. The language evaluation revealed spoken English that was almost unintelligible due to interference from Spanish grammar and pronunciation, and a very limited vocabulary outside of medical terminology. In addition, Dr. R reported difficulty in comprehending the spoken language of African American patients, colleagues, and staff.

The short-term English for medical purposes (EMP) program was designed to address the tasks that Dr. R needed to be able to perform in order to return to work: conducting medical interviews with patients, presenting cases to the supervising physician, and writing and dictating discharge summaries. Because Dr. R's supervisor had identified her as someone needing language support, she was released from her responsibilities on the following schedule. In a 4-day cycle, she was on call the first day, on postcall (and consequently sleep deprived) the second day, and released for 1 hour of language instruction on the third and fourth days. This amount of time is appallingly small considering the nature of her difficulties and the daily work demands placed on her. Needless to say, it was not optimal for language learning, but it helped give her the support she needed to survive the month. After that month was complete, the program director gave Dr. R lighter rotations for a few months to allow her to concentrate on her language skills and to give her some emotional support.

During each of my sessions with Dr. R, a series of role plays based on Dr. R's patients served as the classroom material for developing skills for interviewing patients and presenting cases to supervising faculty. Dr. R identified a patient that she had seen over the past few days. She explained the case with sufficient detail that I could play the role of the patient, and together we role-played the medical interview. In an ideal situation, the role play might have been audiotaped, and Dr. R could have transcribed the tape to improve her ability to self-monitor. In this case, however, as in many ESP settings, time was a significant determinant of methodology. Fortunately, I had had enough experience in ESL and medical interviewing to be able to take notes during the interview regarding mistakes or suggestions for improvement. After performing the role play and discussing the patterns of errors, I gave suggestions for improvement, and we repeated the role play in order to practice the changes. I kept detailed records of the patterns of errors so that I could systematize the suggestions for improvement and match them to the language needs of the resident; in addition, I could document improvement, refer to previous lessons, and avoid bombarding the resident with random suggestions for improvement.

The next step was another role play. In this case, I played the role of the supervising physician, and Dr. R presented the case to me. This task required the use of technical medical language to give details of the case according to a formal structure, punctuated by interruptions by the "supervisor" to ask clarifying questions. This role play was also followed by suggestions for improvement and a repeat of the role play.

As Dr. R began to show improvement and become more confident, the lessons expanded into a variety of other tasks, such as answering pages on the telephone, discussing diagnostic results with family members, transcribing videotapes of African American patients, and writing abstracts for submission to professional journals. Dr. R continued to rely on me for help when she needed it, but as her language skills improved, her confidence increased, and her intelligence, medical knowledge, and compensatory skills began to emerge.

Dr. R's case illustrates not only the challenges of teaching EMP but also the principles of teaching ESP. The curriculum evolves through a constant interaction among the teacher, the student, the language, and the tasks to be performed. The instructor needs to have familiarity with the tasks and, more important, flexibility in teaching and curriculum development as well as compassion for the professional as an individual with dignity, intelligence, and emotional needs.

◈ DISTINGUISHING FEATURES

A Full-Time Instructor

In Wayne State's language and communication skills program for IMGs, the presence of a full-time ESP/communication skills instructor who offers a program that is supported by the department and integrated into the residency training program is unprecedented. Many residency programs offer accent training through local speech pathologists or ESL remediation through local universities. However, the fast pace of the residency program, the high stakes of failure, and the highly technical nature of the communication tasks require a credible instructor who is immersed in the environment, very flexible in terms of time, and creative in curriculum delivery. Needs analyses and instruction take place in a variety of environments—at times in a quiet office, but at other times at a patient's bedside, in the clinic treatment room, or at a medical lecture. I read medical journals and attend conferences on health communication and medical education in addition to keeping abreast of developments in ESL and ESP.

Naturally, most medical residency programs cannot afford a full-time instructor; neither can the Wayne State University program. Therefore, the position is funded not solely for ESP instruction but also for general communication skills for the residents and ESP instruction in the adjacent doctoral programs in the medical research departments (e.g., physiology and immunology). In this environment, what is essential is not a medical background but a keen interest in the field of medicine and a flexible and creative spirit.

The Instructor's Authority

The second distinctive feature of the program is related to the first in that it would be difficult to achieve without a full-time professional. As a full-time faculty member, I have the authority and the confidence to shadow residents in their professional setting, videotape them with their patients, assist them in their communication tasks, and, when necessary, require them to get the help they need. The presence of a communication professional with authority sends an important message: that the residency training program cares about its residents, about physician-patient communication, and ultimately about patients.

❖ PRACTICAL IDEAS

Offer Practical Solutions to Program Directors

Residency training program directors and medical educators often realize that the IMGs in their programs need a special program to help them develop language and cultural skills necessary to perform their daily tasks successfully. However, they are generally unaware that ESP is a professional field with competent, well-educated professionals who can serve as consultants or faculty.

ESP instructors might begin by meeting with residency program directors and offering to demonstrate some practical ways to help IMGs adjust. For example, conducting individual language assessments or holding group or individual reviews of videotaped interviews with real patients can build the ESP instructor's credibility and provide ample material for classroom instruction. In addition, the ESP instructor can offer to assist residents in many small but significant ways, such as preparing professional presentations, simulating job interviews, writing curriculum vitae and personal statements for employment, and providing on-the-spot language assessments.

Tailor the Instruction Schedule to the Residents' Schedules

IMGs are generally eager for ESP support, but the allocation of funds is only the initial step. Residents need to be released from their busy schedules to concentrate on developing language and culture skills. One possibility is to teach month-long courses to small groups or individuals during noncall or elective months; another is to teach on evenings or Saturdays, when the residents can get away from the hospital setting. Yet another is to shadow individual residents in their daily work, providing focused or on-the-spot feedback. For example, the ESP instructor might observe a resident interviewing a patient in the clinic or the hospital and then have the resident present the case to the instructor and receive feedback before presenting it to the medical faculty. If this shadowing is done in a supportive rather than a threatening environment, the resident will benefit greatly and the instructor will gain new insights into the needs of the resident.

Become Comfortable in the Medical Setting

Another key is for the ESP instructor to be well prepared to function in the medical setting. The instructor does not need a medical background to gain this expertise. Rather, it can be gained by reading medical journals and medical student textbooks, watching medical dramas on television, observing doctor-patient communication, and keeping abreast of developments in the fields of ESP and EMP through journals, conferences, and textbooks. In addition, journals devoted to health communication in the fields of medicine and communication studies can provide more theoretical and practical background information.

◈ CONCLUSION

Although I had no background in medicine, my work as the ESP instructor in a language, culture, and communication skills program for IMGs in an internal medicine residency training became an integral part of the medical training, providing training in culture, interviewing skills, and language skills to international and U.S. medical residents. The communication skills curriculum is limited in two important ways, both of which are typical of on-site ESP training programs. First, although I am fortunate to have the endorsement and support of the administrators and faculty in the residency training program, I generally work alone and cannot possibly meet the diverse language, culture, and communication needs of more than 100 medical residents. Second, the residents are under tremendous pressure to perform at a consistently high level in an environment that is physically, intellectually, and emotionally challenging; instruction in communication is sometimes considered frivolous and extraneous. At the same time, however, the addition of the Clinical Skills Assessment as a requirement for incoming IMGs and the greater emphasis placed by the certifying board in internal medicine on communication skills as an integral part of residency training mean that the role of the ESP professional has become increasingly relevant.

This program shows how ESP curricula and materials are in a constantly evolving, interactive relationship with the communication context and the needs of the learners. Additionally, it illustrates the significant contribution of ESP professionals in a highly technical, high-stakes environment. By far the most important contribution of this program is that it improves the quality of communication between physicians and their patients and colleagues and therefore ultimately benefits both physicians who care about their patients and the patients who seek to improve health through a relationship with a caring and compassionate physician.

◈ CONTRIBUTOR

Susan Eggly is an assistant professor in the Department of Internal Medicine in the School of Medicine at Wayne State University, Detroit, Michigan, in the United States. She taught English for academic purposes for 12 years before moving to the School of Medicine to develop a language and culture curriculum for IMGs and a course in oral proficiency in English for doctoral students in the basic medical sciences. Her research has focused on language and culture in a medical setting and on other aspects of physician-patient communication, and has been published in both medical and ESL journals. She has provided teacher education in English for medical purposes throughout the United States and internationally.

CHAPTER 8

An ESP Program for Management in the Horse-Racing Business

Rob Baxter, Tim Boswood, and Anne Peirson-Smith

❖ INTRODUCTION

Writing for Committees was a collaborative ESP/management communication project designed and run by the Department of English at the City University of Hong Kong (the consultants, which included two of us, Boswood and Peirson-Smith) for the Training Department of the Hong Kong Jockey Club (the client, for which one of us, Baxter, was the in-house trainer). The project aimed at developing the capacity of senior managers to write committee papers, the documents that drive top-level corporate decision making in the organization.

Writing for Committees serves as a good illustration of standard ESP methodologies (Robinson, 1991) but goes beyond these in several important respects. We followed best practice in ESP in that

- Program design was based on an in-depth needs analysis conducted through corpus analysis, questionnaires, and interviews with senior managers and potential participants in the training.

- The content of the training materials related to the specific activities of the participants.

- The training program focused on language pertinent to organizational activities.

- Committee papers, the focal documents, are a highly specific target genre with a critical corporate role.

- Training was oriented around case studies, including video and printed input, supported by microplanning and group and individual writing activities.

- All materials were written and produced specifically for the client.

- The program received intensive, multilevel summative and formative evaluations from senior managers and participants that helped refine the program to meet the organizations' needs more closely.

The program also presented several distinctive challenges and opportunities because

- The participants were a demanding group of experienced and highly qualified senior managers from a wide variety of professional disciplines.

- Most of the participants were highly competent in the use of spoken and written English as their second language, and a minority of the participants were native English speakers.

- The program was an innovative, high-profile course for the corporation and, as such, involved a degree of risk, demanding close collaboration and trust between client and consultants.

- We employed a variety of delivery modes, including seminars and individualized coaching to maximize the transfer of learning.

- In addition to individual skills training, we developed a computer-based document template, a printed style guide, and guidelines for managing collaborative writing processes.

- This was a collaborative contract. The consultants designed programs with the client and turned the course over to in-house trainers, who observed a sample of training sessions throughout the delivery period.

The training program can most accurately be classified as English for specific business purposes (Dudley-Evans & St. John, 1998, p. 56) given the specialized nature of the course content and the need to teach the strategic English language skills required to plan and write a single genre, the committee paper, within an organizational context. The program has since won two awards, in Hong Kong and the United States: the Hong Kong Management Association's 1998 Excellence in Training award and the American Society of Training and Development's 1998 Excellence in Practice award.

◈ CONTEXT

The Hong Kong Jockey Club

The Hong Kong Jockey Club is a major Hong Kong not-for-profit organization with betting turnover approaching HK$90,000 million per annum (U.S.$12 billion). Though established as a company, the club has no shareholders and is overseen by a board of 12 unpaid stewards, who are leading figures on Hong Kong's commercial and political scene. The Hong Kong Jockey Club plays an important role in the life of Hong Kong. It operates all horse racing and betting and, on behalf of the government, the Mark Six lottery. The club is also well known for its community and charity work. As a not-for-profit organization, after covering operating costs and payment of tax it donates all surpluses (amounting to approximately HK$1,000 million [U.S.$129 million] per annum) to Hong Kong community projects and charities. It employs almost 5,000 full-time and 15,000 part-time staff.

Participants were drawn from the top 300 executives of the club. The target group consisted of 42 senior and 24 junior executives, the core group responsible for producing committee papers. These executives, most of whom were native speakers of Cantonese, were midcareer professionals with advanced or native competence in English. Their disciplines varied across a wide range of corporate activities, including financial management, construction project management, computer engineering, information systems development, software engineering, human resources management, veterinary medicine, racing management, catering management, community outreach, and corporate development.

Problem Identification

Like that of many other large Hong Kong–based organizations, the club's communication culture is based very much on writing. The decision-making process is leveraged through the production of documents such as committee papers, their discussion at various levels in the organization, their amendment, and their final approval by senior management. The highest level of decision making is the Board of Stewards and its various committees and subcommittees. Through committee papers, the club's management presents proposals for major projects (and their funding), for policy changes, and for the granting of donations to charities and community projects.

The program originated in late 1995 when a number of factors, both external and internal to the organization, focused attention on executive-level writing skills and on committee papers in particular. The general understanding was that, as a conduit for informed decision making at all levels of the Jockey Club, these papers should be written according to the highest professional standards regarding information, argumentation, tone, and style.

At the time of this ESP project, the club had begun embarking on a major change aimed at achieving the corporate mission of "Total Customer Satisfaction" by enhancing or achieving six corporate goals. This process (still in progress at the time of writing) involved substantial reengineering of the organization (including the adoption of a performance management system) and in particular a more rigorous process for assessing and approving projects. This change put more pressure on executives to produce high-quality writing and, in particular, more detailed and argued cost and market-based analysis. As one of several executive development programs, the ESP program was seen as a step in achieving Total Customer Satisfaction (see Figure 1).

The chief executive and directors became aware that a skills gap could develop with the departure (or imminent departure) of those executives traditionally responsible for writing papers. With no systematic means of transfer to the next generation of executive paper writers, the club risked losing the knowledge and skills

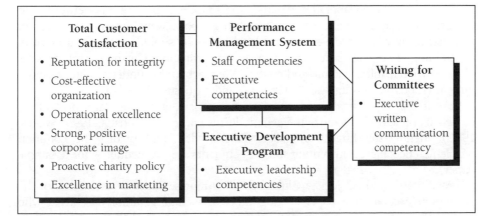

FIGURE 1. Alignment of Program With the Organization's Mission, Objectives, and Strategies

base of the departing executives. Indeed, these problems were already becoming manifest, notably to the target audience—the stewards of the Jockey Club. Draft papers produced by executives were excessively long, were poorly structured, and contained irrelevant operational detail. Getting the drafts into a condition fit for presentation to committees was requiring increasing amounts of editing time on the part of senior managers (particularly directors)—time that could clearly be better devoted to running the business.

In the runup to the 1997 handover of Hong Kong to China, a lot of attention turned to developing *Putonghua* (i.e., Mandarin Chinese) skills in preparation for communicating in the third official language of the future government. Although the club was also running Putonghua programs, club executives felt that such programs should not run at the expense of the language skills executives needed to function in an organization where English was the predominant medium of internal written communication and where links with the international racing community were growing in importance.

◈ DESCRIPTION

Needs Analysis

Methods

The needs analysis began with a preliminary discussion among directors, conducted by one of us (Baxter) as to needs and expectations. This step was important. Directors, as the ultimate gatekeepers of paper quality, were the key stakeholders, so satisfying them would become the ultimate measure of success. They were also integral to the drafting and writing of committee papers and gave critical insight into the key problem areas to be addressed, possibly through a training program.

Following approval of the project, consultants from City University of Hong Kong were appointed, among them two of us (Boswood and Peirson-Smith). We conducted a fuller needs analysis to ascertain the precise training requirements. The intention was to gain a deeper understanding of the potential participants' writing skills, attitudes toward writing papers, and personal characteristics affecting their learning styles, coupled with the needs and expectations of their target readers and gatekeepers in the collaborative report-writing process. In addition, we needed to explore the organizational context in which the reports were written as a means of understanding which professional and language skills should be the focal point of the training in the interests of best practice and compatibility with the requirements of the corporate infrastructure and culture. We reached this understanding through three methods: corpus analysis, interviews, and a questionnaire survey.

An initial corpus analysis of 30 committee papers, at various stages of drafting, helped us broadly characterize the generic forms and moves (Bhatia, 1993; Swales, 1990), identify the English language skills needed, and predict the common problems encountered in writing these documents (see Appendix A for a sample paper).

We interviewed 20 Jockey Club executives representative of each stage of the collaborative committee paper–writing process, including junior executives who drafted the papers, senior managers who edited and polished them, and directors who acted as gatekeepers ensuring that the content and presentation of the paper conformed to the expectations of the target readers—the Board of Stewards. The semistructured format of the 1-hour interviews enabled us to explore key themes such as the organizational process of writing committee papers and the personal experiences of those involved in this activity with some consistency across the sample of respondents while allowing for the possibility of exploring issues that the interviewer had not anticipated. Typical questions included

- What is your role in the Hong Kong Jockey Club?
- What types of writing does your job require?
- What is your role in the committee paper writing process?
- How long have you been involved in writing committee papers?
- How many people are involved in the process of writing committee papers?
- How often do you write committee papers?
- How long does it take on average to write a committee paper?
- How much of your time is devoted to writing committee papers?
- How do you feel about writing committee papers? (Prompt: Is it a positive or negative experience?)
- Who do you write committee papers for, that is, who is your target audience? What type of feedback do you receive from them about these papers?
- Do you feel qualified to write committee papers?
- Which are the easiest aspects of writing a committee paper?
- What are the hardest parts of writing a committee paper?
- How could you improve on your committee paper writing skills/ processes?
- Would a training program covering how to write committee papers be useful to you? Why/why not?
- What aspects of committee paper writing should the training program cover?

Usually, the in-house trainer (Baxter) was present at the interviews as participant and observer. Some of the interviews were tape-recorded, which enabled us to develop a more natural flow of conversation without having to concentrate on note-taking. In this way, we could build up a comprehensive picture of the potential participants' writing competencies and training requirements within the organizational context of the Jockey Club. Before attending the training, all participants also completed a questionnaire survey (see Appendix B) consisting of 16 closed- and open-ended questions covering issues of time management involved in the writing process, attitudes toward writing committee papers, approaches to writing the first draft, an evaluation of strengths and weaknesses when writing, identification of key writing skills, and expectations for the training program.

The function of the questionnaire was twofold. First, it provided detailed data about the executive writers' perceived role in the process of writing committee papers, their perceived writing ability, and their attitudes toward this writing task. This information was useful in ensuring that the training materials matched the realities of the participants' involvement in the management and writing of committee papers. Secondly, the questionnaire helped the participants focus on their involvement in writing committee papers from the viewpoint of strategic management as well as personal skills. This information would then act as a competency-tracking device, enabling both participant and trainer to review the questionnaire responses of each participant at the first coaching session to determine any changes in perceptions about writing committee papers and skills transfer following the training program.

Findings

Interestingly, the needs analysis revealed that the problem could not be explained simply in terms of individual competencies and training requirements (i.e., a training gap) but involved systemic development. This discovery moved the project toward a performance development approach as described in the work of Robinson and Robinson (1989, 1995, 1998). Following this approach, we were intent on developing streamlined systems for the collaborative writing of committee papers (i.e., development of the document cycling and approval system through hard and soft technology) as well as individual skills in composition.

Unsurprisingly, given the caliber of executives involved, the needs analysis showed that all had an excellent grasp of the writing fundamentals (grammar and style). Two problems, however, were apparent:

1. The papers were not always effectively structured and argued (it took time for the reader to find the main argument) and were often inconsistent in format and overly long.

2. The drafting process that preceded the submission of papers to committees was lengthy and inefficient. Typically, draft papers were long and contained irrelevant information. Rather than presenting a strategic-level case using the kind of persuasive argument techniques necessary to obtain senior management's approval for policy change or funding, these drafts focused on merely informing at an operational level of detail. In short, writers were telling, not selling.

As a result, executive writers, or their more senior managing editors (often up to the director level), had to spend excessive time editing, removing irrelevant information, and reshaping the argument structure. This redrafting became a cause of considerable resentment on the part of both writers and their seniors. One interviewee reported a paper being redrafted 16 times. The sources of these problems were as follows:

- Executive writers did not have a clear idea of the correct format of committee papers. Existing standards were neither fully comprehensive nor widely disseminated.

- Executive writers were often unclear about the requirements of their target readers, the stewards, resulting in a lack of focus and direction in the papers. They were often unsure whether they should write the paper for their line manager, the director or chief executive, or the Board of Stewards.

- Executive writers, focusing at the operational level, lacked the skills and awareness to obtain and define strategic-level document content and construct persuasive arguments.

- Executives in their capacity as writers, and often also when acting as managing editors, lacked the skills to manage the writing process. As writers they were not thinking strategically and systematically about the process of gathering information and drafting papers; as managing editors they were not providing sufficient guidance on the positioning and drafting of papers. In what was effectively a complex collaborative writing process involving multiple writers, there was notable miscommunication (see Figure 2).

Program Objectives

This analysis led to the formulation of the three objectives for the program:

1. educate executives about the particular features and functions of committee papers

2. improve the strategic content and persuasive argument structure of the committee paper itself

3. help executives manage the process of writing committee papers more efficiently in two respects:

   - as paper writers, to think more strategically and systematically about the process of gathering information and drafting papers

   - as managers of other writers, to give guidance on the positioning and drafting of papers

The general intention of the training program was to achieve a sustained, long-term improvement in the writing of committee papers. In other words, the approach advocated in the training course would become the established way of writing committee papers throughout the organization. We would achieve this goal by

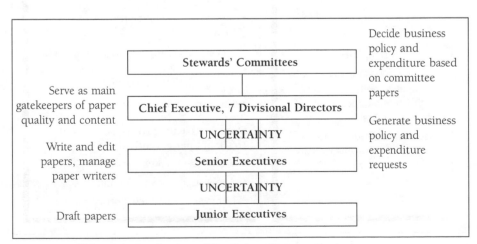

FIGURE 2. Collaborative Production of Committee Papers

creating tangible guidelines such as a style guide and on-line committee paper templates and reinforce it in individual follow-up coaching sessions. We presented our findings an analysis of strengths, weaknesses, opportunities, and threats (SWOT) and a proposal for intervention.

Training Strategy

Given the needs and the resultant objectives identified in the initial investigation, a traditional training solution—the classroom-based, product-focused course—was clearly not appropriate. Rather, we devised a dual solution, involving a style guide and a training program, basing training on systemic development (see Figure 3).

For the training program, two of us (Boswood and Peirson-Smith) team-taught a 2-day, in-house seminar entitled Writing for Committees: An Executive Training Program in Persuasive Writing and provided three individual 90-minute follow-up coaching sessions for each participant. The SWOT analysis had highlighted two main areas to focus on in the training—individual skills in planning and composing the committee papers (the textual aspect), and collaborative skills in efficient management of the writing process (the process aspect). Essentially, these two areas of need served as the basis for the course design.

The training seminars, each comprising 8–15 participants, were oriented around a range of learning experiences maximizing the experiential nature of the course and facilitating the application of knowledge, skills, and theory. These activities included case study scenarios presented through video and printed input, group case analysis and paper planning, team presentations, microwriting practice, and extensive drafting of papers. Course materials included bound handouts and staged exercises

---

Committee Paper Style Guide

- Lays out new standards for papers endorsed by Board of Management
- Serves as tangible means to sustain consistency over the long term

Templates

- Organizationwide networking facilitates speedy and consistent production of papers on PCs.
- Executives (during course) and executive secretaries are trained to use them.

Two-Stage Training Program

Stage 1: Two-Day Training Workshop

- Management of the writing process (video-based behavior modeling and role play)
- Models of committee paper structures
- Methodology of gathering relevant content, planning, writing, and editing
- Simulation of writing using two organizational case studies
- Computer template training

Stage 2: Coaching (three sessions of 1½ hours per executive)

- Feedback on course writing
- Application to own committee papers
- Competency profile/report at the end

FIGURE 3. Systemic and Training Interventions

prepared specifically for the training program, featuring extracts from authentic committee papers that exemplified both good and bad practice. Our role was to present the key concepts, direct discussions, give input as needed during the individual and group writing exercises, and ensure continuity across the development of two committee paper-writing tasks through the coaching sessions, giving feedback on writing competencies of participants throughout.

During the 2 days we presented the conceptual framework, based on a series of models and diagramed argument forms advocating best practice in writing committee papers. The participants applied this framework in critiquing and writing two committee papers. The conceptual framework was introduced to the participants in stages, enabling them to absorb the key ideas and apply them through task-based, practical writing exercises.

Day 1: Managing the Writing Process

The first day of the training course presented techniques for managing the writing processes and preparing papers using problem-solution structures.

The opening sessions focused on managing the individual and collaborative writing processes involved in creating committee papers, the intention being to increase the efficiency of the writing process overall. Initially, participants reflected on the problems they faced as writers and shared them with the group. We collated the responses, matching them with the interview findings addressing the same issues. Typical problems cited included getting started, arguing persuasively, writing quickly, selecting and organizing relevant information, writing with a good style, understanding superiors' requirements, and writing with a clear focus. This reflection led the participants to acknowledge the personal and team challenges faced in writing papers and recognize that such challenges were universal, were related to corporate culture and teamwork processes, and, most importantly, would be addressed in the training program. To crystallize the participants' thoughts and demonstrate the immediate relevance of the course to their writing tasks, we presented a worst-practice committee paper for collective comment and comparison with a best-practice paper. In this way, the participants had benchmarks to follow and refer back to throughout the training sessions.

We then introduced the conceptual framework characterizing good and bad practice in the writing of committee papers. Materials included a simple model of the writing process (Figure 4) and the People-Results-Positioning-Information-Argument (PRePIA) model (Figure 5) for planning effective papers. The PRePIA model served as a graphic overview of the recommended approach to planning persuasive papers, advocating that the writers position themselves strategically for effective writing by identifying the problem, the readers, and the intended results; clarifying the information readers needed; and devising a persuasive argument that delivered a clear message.

The participants then applied this planning process to the task of positioning a committee paper. They viewed a video re-creation of a typical briefing meeting between a junior and a senior manager consisting of two scenarios—best- and worst-practice briefings—and noted from their observations why the briefings would have resulted in a good and a bad paper, respectively.

The participants next examined a series of models representing the internal textual structure in preparation for writing their first committee paper. Overview

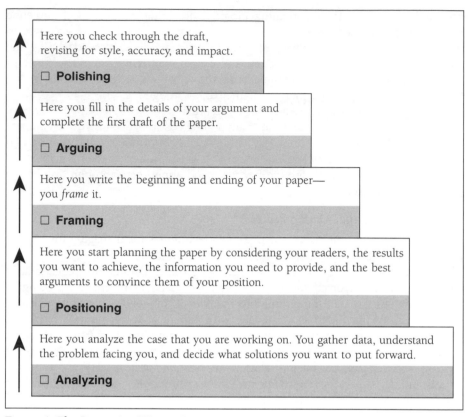

Here you check through the draft,
revising for style, accuracy, and impact.

☐ **Polishing**

Here you fill in the details of your argument and
complete the first draft of the paper.

☐ **Arguing**

Here you write the beginning and ending of your paper—
you *frame* it.

☐ **Framing**

Here you start planning the paper by considering your readers, the results
you want to achieve, the information you need to provide, and the best
arguments to convince them of your position.

☐ **Positioning**

Here you analyze the case that you are working on. You gather data, understand
the problem facing you, and decide what solutions you want to put forward.

☐ **Analyzing**

FIGURE 4. The Persuasive Writing Process

models included the organizational diamond (title—aims—information and argument—recommendations) and the attention—needs—solutions—action (ANSA) approach to formulating persuasive argument. More detailed models included alternatives for single- and multiple-option papers (with checklists; see Appendix C), derived from the basic information frame given in Figure 6. These models were adaptations of structures developed by Jordan (1984) (situation—problem—solution—evaluation) and Blicq (1993) (summary—background—facts—outcome).

Throughout the presentation of conceptual models, the participants examined working examples taken from a committee paper, with extracts often rewritten to highlight particular points. They also completed writing tasks at the end of each stage to enable them to practice the newly introduced concepts from the basic information frame. Once participants had familiarized themselves with the major concepts, they reviewed in groups the notes they had made in positioning the best-practice paper and used both the organizational diamond and the ANSA model to draft the structure and persuasive argument of their paper individually using the computer-based templates provided. As the participants worked on their papers, we circulated, giving individual guidance. By the end of Day 1, the participants had each produced the first draft of their committee paper, which we reviewed and edited overnight using the checklists (see Appendix C), providing detailed comments and a model

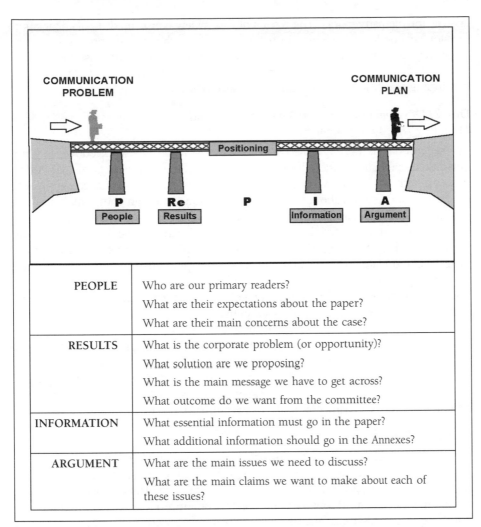

FIGURE 5. The PRePIA Model for Preparing a Communication Strategy

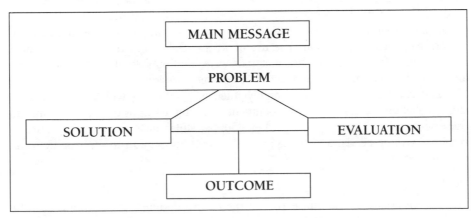

FIGURE 6. The Basic Information Frame

answer for the participants. The results of this review constituted the feedback and revision session for the opening of the second day of the training program.

Overnight, the participants were asked to read background case notes for a second committee paper to be written during Day 2 of the training program, continuing the theme of simulating workplace practice.

Day 2: Developing Individual Writing Skills and Modeling the Management Process

The second day of training aimed at enhancing the ability of the executives to write persuasive committee papers by focusing on the microlevel aspects of the writing process, such as tone, style, coherence, and clarity. Continuing the staged approach, the participants learned about these writing techniques through course materials and additional working examples that we presented.

The second day's training centered on the second case study, which the participants had prepared overnight. The participants reviewed the case study materials, and planned (in teams) and wrote up (individually) a committee paper, applying all the concepts and models introduced throughout the course up to that point. In a simulated briefing session, one of us acted as controller (senior manager), and the participants' teams acted as executives. The executive teams were encouraged to plan their paper by asking the controller relevant questions as a precursor to positioning the document according to the PRePIA model. Then, using an argument and style checklist (see Appendix C), participants then peer-reviewed the completed first draft of this second committee paper in pairs, focusing on micro aspects of writing style. One of us reviewed this paper further with each participant in the first coaching session, ensuring continuity between the two stages of the training.

Coaching Sessions

One-to-one coaching sessions took place in participants' offices and involved working, whenever possible, on documents that were currently in progress. Often we worked with a participant on a paper throughout the three coaching sessions, taking the document through its life cycle from inception to final editing and polishing for submission to the director or Board of Stewards, depending on the role of the participant. When no suitable current writing task was available, the coaching session focused on either a case from the manager's files or a specially developed case study. Also, we used writing tasks drawn from the course materials (which the participants were asked to prepare as homework) as warmup revision exercises at the start of each coaching session. The use of writing tasks from the course facilitated transfer, ensuring that the course materials were used outside of the training room, and provided continuity among the three coaching sessions.

The coaching sessions were highly intensive, involving wide-ranging discussions, collaborative case analysis, document planning, composition, and editing. In these sessions, each participant could explore particular areas of concern in terms of personal writing needs, and we tailored the coaching session to each participant's requirements.

Evaluation

Summative evaluation, a critical part of the course design, served to optimize the learning process and ensure that the training course was relevant and responsive to the changing needs of the organization. The program was evaluated as follows.

To measure reaction to the workshop, all executives were asked to complete an evaluation survey. One of us (Baxter), as the in-house trainer, also sat in on the course as an observer, giving detailed comments at the end of each course. The other two of us, as the trainers, used this feedback to make adjustments to workshop management and training materials as the course progressed.

To give immediate feedback to participants on their performance at the end of the coaching sessions, we completed a competency assessment profile evaluating the participant's writing proficiency in the areas of argument structure, argument content, style, grammar, and information management (see Appendix D). This profile was a useful way of tracking progress from pre- to posttraining, enabling us to evaluate individuals' competency levels according to predetermined categories as independent writer (mentor status), competent writer (practitioner status), developing writer (apprentice status), and novice writer (novice status). Participants may show progress in some areas of textual competency above others; for example, one might progress from a developing to a competent writer in terms of argument structure while remaining a developing writer in terms of grammatical accuracy.

To evaluate the longer term impact on the organization, we conducted a follow-up evaluation 6 months after the last program by

1. analyzing papers produced by executives in the previous 6 months and comparing them with papers produced by writers before training

2. sending questionnaires to participants following the training program to determine changes in their approach to the management and writing of papers and comparing their responses with those on the precourse questionnaire

3. conducting follow-up interviews with a sample of participants and with the directors to clarify questionnaire feedback and gauge the impact from the perspective of the gatekeepers of committee papers

Table 1 summarizes the results against the three objectives of the program.

◈ DISTINGUISHING FEATURES

Although this project was typical of workplace ESP in many respects, it also presented some special challenges.

Collaboration With Senior Managers

Senior managers are challenging to work with at any time, and in this case the managers represented a wide range of professions. In addition, this program represented the first time the club had offered training in writing to such senior staff, and it was clear from the outset that the program had to be of the highest quality, which influenced the course design and content. The course design was founded on the belief that, to maximize the involvement of these senior executives, the training should replicate reality through two case study scenarios requiring the production of two committee paper drafts over 2 days, involving significant conceptual challenges, collaborative effort, and the individual flair typical of the workplace experience. These pressures also required close collaboration and trust between client and

TABLE 1. MAIN FINDINGS FROM THE PROGRAM EVALUATION

Objective	Result
• To educate executives about the specific features of committee papers	Very significant improvement. Not surprisingly the combination of the training program, style guide, and computer template resulted in very few inconsistencies in document design.
• To improve the strategic content and persuasive argument structure of the written product	Improvement in some significant respects. Executives reported that they were better at selecting relevant content and getting started. The models for strategic planning of the document (PRePIA) and argument structure (organizational diamond) with revision checklists helped writers visualize the intended document from the outset. More significantly, directors reported that they were editing substantially less. Papers, although they did not always adopt the exact models of persuasive structure presented during the course, showed evidence of more persuasive structures. Lengthy background sections have gone, and persuasive argument is more up front. In effect, executives have absorbed the lessons and have been applying them in their own way to the particular writing task they face.
• To help executives manage the process of writing committee papers as paper writers and as managers of other writers	Some sporadic improvement. Some executives have clearly understood the imperative of applying management skills to writing and of collaborating on the writing project and report productive results. For others, however, there has been less of a change in their working relationships.

consultants during the 2-day training program and throughout the individual coaching sessions. The training was a partnership: Participants were treated as respected equals rather than as passive students.

A Document Life-Cycle Approach

Structuring the 2-day training program around the preparation and drafting of two committee papers was productive but required time management. All participants completed two complex documents, which afforded a sense of achievement, especially as one of the major perceived obstacles to writing effective committee papers had been the inordinate amount of time devoted to their planning and execution. In addition, participants applied the conceptual models to plan and create their paper, providing them with the confidence to use the models as a blueprint for their future writing tasks.

Integration With Corporate Values

The program was delivered during a period of intensive corporate restructuring and repositioning. In consequence, several aspects of the training were pertinent to issues of corporate development. Case studies had to be carefully written to avoid internal sensitivities. Argument forms, particularly those of service improvement and prudent financial management, were explicitly linked to the components of the Total Customer Satisfaction strategy.

Individualized Coaching

The push for quality led us to employ individualized coaching sessions, taking training beyond the classroom and into the daily activities of the workplace. This one-to-one teaching approach enabled us to identify the language training needs of individual executives by reviewing their expectations of the training (submitted in the pretraining questionnaire) and evaluating the two committee papers written by the participant during the course. In this informal setting we could respond in detail to the participant by focusing on those aspects of the writing process that the participant found most challenging and explore these at the individual's pace. Equally, in the coaching sessions some participants preferred to focus on papers that they were actually writing while others used a case study for extra practice and supplemented with revision exercises highlighting specific aspects of the course material such as persuasive argument content and structure or style. In dealing with particular problems in executive-level writing, our role during these one-to-one sessions was multifaceted, ranging from sounding board and confidante to coach. This partnership approach was essential in motivating the participant to apply the training materials to everyday work practices.

Participants as Tutors

A significant, unexpected development was that participants, notably senior managers, tutored their subordinates by sharing aspects of the program when briefing them on writing committee papers and other documents. This transfer of knowledge ensured that the training program achieved a critical mass, took on a life independent of the training room, and contributed to the sustained application of the materials. It also underlined the collaborative nature of writing within the organization, which gave further credence to the development of the training materials around both group and individual input.

A Collaborative, Turnkey Project

This program was a highly collaborative, turnkey contract: As consultants and in-house trainer, we developed the programs jointly, and the course was handed over to the client for independent delivery. In a follow-up project for middle management, we worked out this approach more fully through team teaching.

We worked toward creating a total package of services to support executive writing of committee papers. We developed a new document design, integrated this with the club's Executive Information System through a document template, produced a printed style guide and quick reference cards, and included guidelines for managing collaborative writing processes in the training sessions.

◈ PRACTICAL IDEAS

ESP practitioners engaged in corporate communications training may like to consider some of these ideas in planning their own interventions.

Target a Particular Group or Document Type

Following standard ESP methodology, whenever possible, really works. Course design is easier, the approach is highly focused, participants are more satisfied, and better and more measurable results are likely when a particular group or document type is the subject of the training.

Simulate the Writing Process

Program developers should take full account of the collaborative aspects of the way documents are created in the organization. Documents with organizationwide significance are not generally individual products. When the target documents are individually composed and polished, writing tasks should lead up to individual practice and competence. When documents are collaboratively composed or hierarchically screened, these processes should be built into the training. The participants should proceed through a staged process of writing from case analysis to the positioning, drafting, screening, editing, and polishing of the document.

Offer Training in Managing the Writing Process

Managing the writing of subordinates or peers is a complex and sensitive process, yet it is seldom the focus of management training. ESP practitioners should tap into the insights of practitioners, develop materials and tasks, and supply user-friendly checklists for application on the job.

Link the Training to Corporate Development

Linking the training program to critical issues in corporate development enables participants to relate specific communication strategies to broad management strategies, increasing the motivational levels of the participants. But such issues can raise distracting debate in the training room, especially if the issues are controversial or sensitive.

Position the Program Carefully Within the Organization

ESP practitioners need to sell the importance of good writing skills and get the support of senior management. Top-level endorsement will gain the necessary legitimacy for the training program throughout the organization.

Develop Additional Value-Added Services

Taking opportunities to develop the writing infrastructure—through document redesign, templates, style guides, and quick reference notes—will sustain the training beyond the training room. Does an organizational style guide currently exist? Should it be updated or rewritten? If not, should a new style guide be written? What is the best way of disseminating the guide throughout the organization to encourage consistent usage? Would an interactive computer-based document template speed up the document drafting process and ensure uniformity in terms of document presentation? Can a template be internally developed (and maintained)? Are there any other parties (e.g., secretarial staff) who need training in subskills?

Use In-House Expertise

Using in-house expertise for both the subject area and the training function will ensure the best fit between organizational needs and training solutions. Leading managers can often act as mentors and role models for their subordinates by transferring the knowledge gained during the training. Bringing them into the training room to validate the content will strengthen the program.

Use Executive Coaching If Resources Permit

Many senior executives welcome one-to-one (or one-to-two) coaching. It recognizes the individual differences in their communicative roles, allows confidential discussion of sensitive cases, and increases the chances of successful workplace transfer and sustainability of the training.

◈ CONCLUSION

In this project, the ESP function expanded into the areas of management develop ment because we recognized the strategic role of corporate communication and the persuasive nature of many business documents. In this situation, the Jockey Club was committed to fostering best practice in report writing at the committee level, as these reports were the conduits for informed, strategic decision making at all levels of the organization.

The intensity of corporate change emphasizes that ESP must not be content just to reproduce existing forms of communication; it must contribute to the development of new forms to meet the challenges facing corporate writers operating in a constantly changing workplace. ESP has to engage with, and contribute constructively to, the processes of corporate development, emphasizing the strategic role of written communication and its persuasive function in the modern organization.

We believe that training in writing cannot be conducted independently of the systems within which the writing takes place. To achieve results, both must be brought under scrutiny. This belief broadens the conception of needs analysis from its individualist focus, based on the outdated model of the lone writer, toward communication auditing, which situates individuals within work groups, or communities of practice, that function within the organization as a systemic whole.

Adopting a systems perspective requires balancing and integrating process and product approaches. This project set out to enhance the ability of executives to write persuasive committee papers and increase the efficiency of the overall writing process. Achieving this goal involved improving both individual skills in planning, composing, and editing, and collective interactional skills in managing the writing process. Consequently, the course materials and activities alternately focused on text as product and on interaction as contributing process.

Such interventions demand intensive client-consultant collaborative development.

As consultants and in-house trainer, we worked closely together from the start, providing an insider's view of the client organization, which greatly assisted in determining the precise nature of the training needs and the exact challenges faced by the participants in writing and managing committee papers.

◈ CONTRIBUTORS

Rob Baxter is corporate affairs manager at the Hong Kong Jockey Club. At the time of the project he was responsible for the club's English communications standards, training, and development. He has more than 10 years of experience in the field of ESP in Hong Kong and China.

Tim Boswood is associate head of the Department of English at the City University of Hong Kong. He teaches management communication, ESP program management, technical writing, and desktop publishing. His current research interests include identity issues in management writing and the role of communication training in corporate development.

Anne Peirson-Smith is an assistant professor in the Department of English at the City University of Hong Kong. She teaches promotional discourse, public relations, writing for the media, and advertising copy writing. Her current research interests include persuasive communication in professional and social contexts.

◈ APPENDIX A: SAMPLE COMMITTEE PAPER

Computer Equipment Upgrade for Production, Development, and Testing Systems

Aim

1. This paper asks the Committee to authorise expenditure of $[...] for acquisition of computer equipment to upgrade a number of areas of the Club's production, development and testing systems. Funds were adopted in the 1996/97 capital budget for this purpose.

Background

2. The Club's installed computer base allows for expansion as usage increases. From time to time, installed equipment is supplemented to cope with changing processing demands, additional applications or improved security.

3. This year, additional computer equipment is required to:
   - Upgrade the corporate systems to support new business needs;
   - Implement disaster recovery recommendations to better provide for business continuity in the event of a disaster;
   - Provide secure external network access;

- Upgrade the development and testing systems to improve development and testing of applications.

4. The cost breakdown of these requirements is summarised in Annex A.

Upgrade of Club Information System (CIS) and Corporate Systems

5. The Club Information System (CIS) that runs the financial, administrative and racing applications is being phased out gradually. New solutions based on Open System standards allow the Club to choose products from a wide selection of hardware and software vendors. The Financial Budgeting System, the OCB Roster and Attendance System, the Catering Attendance System, the Help Desk System, the Computer Equipment Information System, and the Staff Social Club Management System are new applications requiring additional equipment at a cost of $[...].

6. The Club has office automation systems supporting documentation, filing, spreadsheets, electronic mail and information sharing and dissemination. With an increase in the number of users, increases in transaction volume and the introduction of new facilities such as the ClubInfonet, the computer equipment supporting office automation is reaching its capacity. Two additional servers are required at a cost of $[...].

Implementation of Disaster Recovery Recommendations

7. A recent review of the IT facilities' disaster recovery capability has identified some deficiencies in the backup Betting Operations Control Centre at Sha Tin. $[...] is needed for it to fully back up the Happy Valley Betting Operations Control Centre in the event of a major disaster.

8. The Media Communication System was introduced to provide racing and betting information to the press. As no backup system was provided in the low cost pilot system, the service will not be available if the computer equipment breaks down. This will adversely affect information dissemination to the newspapers that have embraced the new communication process. $[...] is needed to acquire a backup server to provide system redundancy.

Provision of Secure External Network Access

9. The Internet has established itself as an excellent business tool, particularly in the Racing Division. At present, access to the Internet and external e-mail is only provided to a small number of Club staff. Secure connections are arranged by connecting the personal computer of these users through a separate network. To improve security and expand the service to more users, existing pilot system equipment needs to be upgraded with a commercially available secure firewall that interfaces to the office automation network and prevents unauthorized system access. The cost of a secure firewall arrangement for external e-mail and Internet access to authorized users is $[...].

Upgrade of the Development and Testing Systems

10. To successfully implement reliable systems, effective software development and thorough testing are required. The existing development and testing computer equipment is inadequate to efficiently support present levels of software development, maintenance and testing, particularly in the following systems: Personnel and Payroll, Laboratory Information Management, Customer Input Terminal, Electronic Funds Transfer, Telebet, and Distributed Computing Environment. $[...] is needed to improve the situation.

Recommendation

11. That the Committee APPROVES the expenditure of $[...] to upgrade computer equipment for production, development and testing systems.

Director of Information Technology
JRM/DL/hps

◈ APPENDIX B: NEEDS ANALYSIS QUESTIONNAIRE

Writing for Committees
The Hong Kong Jockey Club

About the Questionnaire

The objective of the Writing for Committees training programme is to develop expertise in the writing of Club committee papers. The programme will focus on the processes involved in writing and the type of documents which are produced. Your responses to this questionnaire will help us to establish your involvement in this process as well as assist you in understanding how strategic Club documents get written and some issues involved in managing this process. Your comments and responses will also be a starting point in the coaching sessions that follow the two-day writing workshop.

We use the term committee paper/report throughout to refer to all types of substantial, internal documents, produced by whatever means, i.e., writing by hand, by word processor, or by dictation.

The questionnaire should take you between fifteen and twenty minutes to complete. Please feel free to write additional comments in the margins if you wish to add more information or if you feel any of the questions are missing the point.

Confidentiality

Please note that your responses will be completely confidential to the trainers. No one from the Club will review these questionnaires, nor will they be made available to any other City University staff. A general report on the outcome of this survey will be made to the Club, but no comments will be attributable to individuals. Anonymity is assured.

Instructions

1. Write your name in the box below.

2. Complete the questionnaire by writing your answers in the blank spaces provided or by ticking in the relevant box.

3. Attach your business card.

4. Return your completed questionnaire in a sealed envelope to Rob Baxter in Training Department.

Name _____

Part 1: Writing in Your Job

In the first part of the questionnaire we would like to find out about the kinds of document you write and how you go about the task.

1. Please indicate whether you, personally, write the following documents as part of your job.

Types of documents	Yes	No
Committee papers in general		
Project progress reports		
Quarterly reports		
Operational plans		
Strategic plans		
Proposals for expenditure		
Proposals for policy changes		

2. In an average week, what proportion of your working time is spent on writing and editing committee papers/reports? _____%

3. Do you feel that this is an efficient use of your time?

Yes, very efficient	
Yes, quite efficient	
I'm not sure	
No, not very efficient (please explain below)	
No, not at all efficient (please explain below)	
It depends (please explain below)	

_____

_____

4. What method(s) do you use when preparing committee papers/reports? (Tick as many as are applicable. "Typing" here includes use of a word processor.)

Write by hand before typing by another person	
Write by hand before typing yourself	
Write directly onto a PC/word processor	
Other (please specify)	

_____

_____

5. Please specify three substantial documents you have worked on recently.

_____

_____

Part 2: The Writing Process

In this part, we would like you to consider how you approach writing committee papers/reports and how you feel about the writing process.

Attitudes to Writing

6. This question includes a number of items which sum up feelings towards writing committee papers/reports. Please indicate, by ticking in the appropriate box, whether you strongly agree, agree, disagree, or strongly disagree with each statement.

	strongly agree	*agree*	*disagree*	*strongly disagree*
a. It's a frustrating task.				
b. I am strategic about planning a writing project (e.g., I schedule the stages a document passes through, the number of drafts, etc.).				
c. I feel confident I understand the objectives of the documents I write.				
d. I panic at the size of the task.				
e. It's difficult to judge how much detail to include.				
f. Writing is an essential contribution to the Club's operations.				
g. I am concerned about finding the time.				
h. I feel concerned about the amount of data gathering required.				
i. I know how to construct convincing arguments.				

j. I don't get enough support and advice.				
k. I am eager to make a start—I enjoy the challenge.				
l. Editing shouldn't be part of a senior manager's job.				
m. A good writer can get it right the first time.				
n. I have a clear idea of the style of writing required.				
o. I feel unsure about who the audience is.				
p. It's excessively time-consuming.				
q. I feel excited—it's an opportunity to impress and/or shape policy.				
r. Writing is a distraction from the project priorities.				
s. I feel excited about data collection.				

Developing the First Draft

7. When writing committee papers/reports, what should be the primary focus of the initial draft? (Tick up to three.)

Document structure		Main message		Facts and information	
Style		Targeting the audience		Positioning	
Grammar		Tone		Other (please specify)	
Argument		Document purpose			

8a. Do you, personally, write first drafts of committee papers/reports?

Yes ☐   No ☐

If No, go to Question 9.

8b. If Yes, tick up to three of these statements which best match your approach to drafting.

I make an outline first.	
I write it like a story with a beginning, a middle, and an ending.	
I start by writing the easiest part first.	
I use a previous paper/report as a guide.	
I start writing with the information I have and then identify the gaps.	
I write the way I think my boss wants it.	
I write the way I think the readers want it.	
I identify the key message I want to get across.	
I specify my purpose in writing first.	

Editing by Superiors

9a. Do you submit your writing to a superior for editing?

Yes ☐ No ☐

If No, go to Question 10.

9b. What aspects of your documents are most often altered during the editing process? (Tick any that apply).

Document structure		Argument		Wordiness	
Style		Tone		Other (please specify)	
Grammar		Information content			

10a. Do you edit the documents of subordinate staff?

Yes ☐ No ☐

If No, go to Question 11.

10b. What aspects of your subordinates' documents do you most often alter during the editing process? (Tick any that apply.)

Document structure		Argument		Wordiness	
Style		Tone		Other (please specify)	
Grammar		Information content			

Part 3: Self-Evaluation

In this part we would like you to consider your own abilities in writing committee papers/reports. Your responses here will form the basis for assessing the changes following the writing workshop and the individual coaching sessions.

11. Please list up to three types of documents which you find most demanding to write and the reason you find them most demanding.

_____

_____

_____

_____

12. What would you describe as your strengths when writing?

_____

_____

_____

_____

13. What would you describe as your weakest areas when writing?

_____

_____

_____

_____

14. Please list any specific types of document which you believe you need to improve upon (e.g., quarterly reports, proposals for changes in policy).

_____

_____

_____

_____

15. In this question we would like you to think about writing in two ways. First, judge the importance of these skills in terms of their impact on the success of the document. Secondly, assess your own ability in this respect.

		Importance				Ability	
		Not im-portant	Impor-tant	Very im-portant	Essen-tial	Low ability	High ability
a.	Getting the document structure right	0	1	2	3	1 2 3	4
b.	Writing to a Club style	0	1	2	3	1 2 3	4
c.	Writing grammatical English	0	1	2	3	1 2 3	4
d.	Constructing persuasive arguments	0	1	2	3	1 2 3	4
e.	Writing concisely	0	1	2	3	1 2 3	4
f.	Linking sections of text	0	1	2	3	1 2 3	4
g.	Creating the appropriate tone	0	1	2	3	1 2 3	4
h.	Selecting supporting information	0	1	2	3	1 2 3	4
i.	Using computer-based writing tools	0	1	2	3	1 2 3	4
j.	Understanding what your superior requires	0	1	2	3	1 2 3	4
k.	Writing quickly	0	1	2	3	1 2 3	4
l.	Managing the writing of subordinates	0	1	2	3	1 2 3	4
m.	Other skills (please specify)						
	_____	0	1	2	3	1 2 3	4
	_____	0	1	2	3	1 2 3	4
	_____	0	1	2	3	1 2 3	4

Expectations

16. What are your expectations for the Writing for Committees training programme?

_____

_____

_____

_____

Thank you for completing this questionnaire. If you have any comments, questions or thoughts on the way Club committee papers/reports are (or should be) written, please note them on a separate sheet and return them with this questionnaire.

For further information about the Writing for Committees training programme, please contact Rob Baxter of Training Department on X7852.

Tim Boswood, CityU Consultants Ltd.
Anne Peirson-Smith, CityU Consultants Ltd.

◈ APPENDIX C: DOCUMENT STRUCTURES AND CHECKLIST

Single-Option Paper

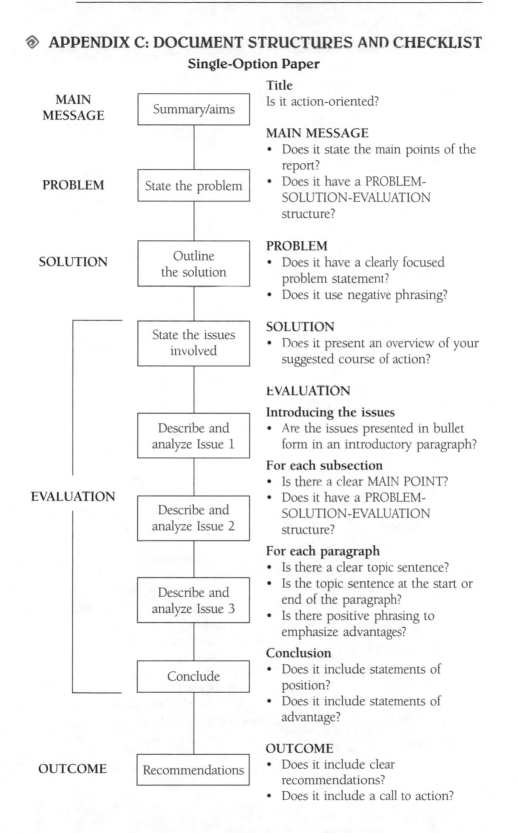

MAIN MESSAGE — Summary/aims

PROBLEM — State the problem

SOLUTION — Outline the solution

State the issues involved

Describe and analyze Issue 1

EVALUATION — Describe and analyze Issue 2

Describe and analyze Issue 3

Conclude

OUTCOME — Recommendations

Title
Is it action-oriented?

MAIN MESSAGE
- Does it state the main points of the report?
- Does it have a PROBLEM-SOLUTION-EVALUATION structure?

PROBLEM
- Does it have a clearly focused problem statement?
- Does it use negative phrasing?

SOLUTION
- Does it present an overview of your suggested course of action?

EVALUATION

Introducing the issues
- Are the issues presented in bullet form in an introductory paragraph?

For each subsection
- Is there a clear MAIN POINT?
- Does it have a PROBLEM-SOLUTION-EVALUATION structure?

For each paragraph
- Is there a clear topic sentence?
- Is the topic sentence at the start or end of the paragraph?
- Is there positive phrasing to emphasize advantages?

Conclusion
- Does it include statements of position?
- Does it include statements of advantage?

OUTCOME
- Does it include clear recommendations?
- Does it include a call to action?

Multiple-Option Paper

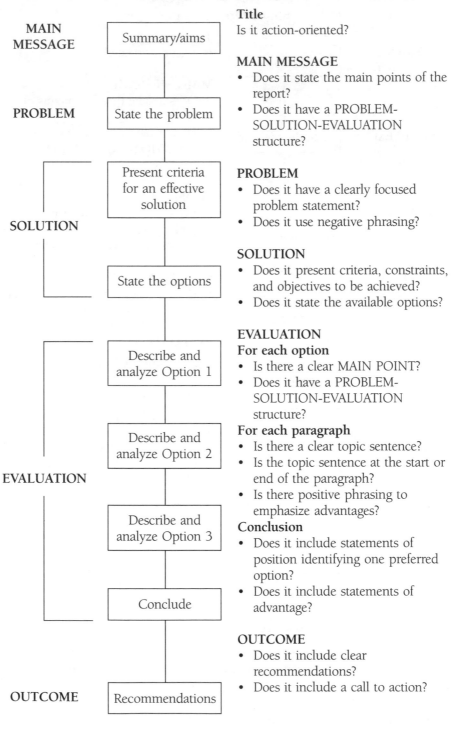

MAIN MESSAGE	Summary/aims
PROBLEM	State the problem
SOLUTION	Present criteria for an effective solution
	State the options
EVALUATION	Describe and analyze Option 1
	Describe and analyze Option 2
	Describe and analyze Option 3
	Conclude
OUTCOME	Recommendations

Title
Is it action-oriented?

MAIN MESSAGE
• Does it state the main points of the report?
• Does it have a PROBLEM-SOLUTION-EVALUATION structure?

PROBLEM
• Does it have a clearly focused problem statement?
• Does it use negative phrasing?

SOLUTION
• Does it present criteria, constraints, and objectives to be achieved?
• Does it state the available options?

EVALUATION
For each option
• Is there a clear MAIN POINT?
• Does it have a PROBLEM-SOLUTION-EVALUATION structure?
For each paragraph
• Is there a clear topic sentence?
• Is the topic sentence at the start or end of the paragraph?
• Is there positive phrasing to emphasize advantages?
Conclusion
• Does it include statements of position identifying one preferred option?
• Does it include statements of advantage?

OUTCOME
• Does it include clear recommendations?
• Does it include a call to action?

Argument and Style Checklist

Evaluate your paper according to the following criteria.

1. Argument

1.1   Does the whole document have a coherent line of reasoning?

1.2   Do subsections/paragraphs make clear claims?

1.3   Are claims sufficiently supported by evidence?

1.4   Are there any logical flaws?

2. Brevity

2.1   Is the paper shorter than three pages?

2.2   Are there any overly long sentences, paragraphs, or subheadings?

2.3   Are there any wordy expressions?

2.4   Are there any *verb+noun* combinations reduced to *verbs*?

3. Directness

3.1   Are there any overly formal expressions?

3.2   Are there too many *-ion* words?

3.3   Are there any "empty words"?

4. Clarity

4.1   Are subparagraphs, bullets, or numbered lists used to clarify complex ideas?

5. Coherence

5.1   Are sentences linked by chaining, adding, or both?

APPENDIX D: COMPETENCY ASSESSMENT PROFILE

Competence → Textual features →	Argument Structure Single- and multiple-option argument structures	Argument Content Problem-solution structures, clear claims and evidence, developed counterarguments	Style Directness, clarity, conciseness, dynamism, coherence	Grammar Sentence-level accuracy	Information Management Information load, distribution of data in paper and annexes, presentation of data
Independent writer (mentor status)	Work has demonstrated independent command of both structures.	Work has evidenced persuasive argument content.	Work has required virtually no editing for style.	Work has required virtually no editing for grammatical accuracy.	Work has required virtually no changes in information content.
Competent writer (practitioner status)	Work has required minor adjustments in argument structure.	Work has required fine-tuning of argument content.	Work has required occasional stylistic improvements.	Work has contained minor surface errors that can be speedily removed.	Work has required minor editing for information management.
Developing writer (apprentice status)	Some sections of the work have required changes to argument structure.	Some sections of the work have required revision of content.	Work has called for substantial stylistic development.	Work has contained noticeable inaccuracies that reduce credibility.	Some sections of the work require adjustment of information content.
Novice writer (novice status)	Work has required major recasting of argument structure.	Work has required extensive recasting of argument content.	Work requires substantial rewriting to reach an appropriate style.	Grammatical errors in work have required extensive editing.	Information content of the work has needed rethinking.

Tick competency levels here.

Definition

Work: the participant's most recent document(s) of the "committee paper" type, taken as representative of the current level of competence.

Course participant:

Assessment made by:

Date:

CHAPTER 9

An ESP Program for Entry-Level Manufacturing Workers

Judith Gordon

◈ INTRODUCTION

ESP training programs for entry-level international workers with low educational backgrounds face the dual challenge of meeting these workers' needs and convincing companies to pay for the training. Yet the number of nonnative speakers of English in the U.S. workforce has grown even beyond earlier predictions that immigrants would make up 22% of new workers in the United States by 2000 (U.S. Department of Education and U.S. Department of Labor, 1988, p. 5). Today many entry-level workers hired by U.S. businesses are nonnative speakers of English (and many speak little or no English). Because the U.S. Census Bureau's prediction that Hispanics would be the largest minority group in the United States by 2005 may have already come true (Larmer, 1999), the need for ESP training programs for such workers will probably continue to grow.

This chapter describes a successful short-term program for training international workers in a U.S. manufacturing company in production-line ESP. The program was offered eight times between 1998 and 2001, and all materials were customized to meet the company's particular training needs and the workers' particular learning needs. Included here are a list of training modules and competencies, a discussion of sample activities, and a description of a multimedia training programming.

The program resulted in increased individual learning, measurable improvements in job performance, and a significant return on investment, in that the monetary benefits of the program far exceeded its costs. These results were due in part to the program's success in creating materials and responding to the learning styles of entry-level workers with limited educational backgrounds.

◈ CONTEXT

The Company

The U.S. manufacturing company in which this program takes place strives to meet all of its clients' quality requirements and even to exceed these requests. The company places great emphasis on producing inexpensive and high-quality products on schedule and thus has been growing quite rapidly. With the aim of gaining recognition as a world-class company, it has applied for certification of its standards by the International Organization for Standardization (ISO 9002), and quality, safety, and advanced production training are offered to all employees.

Very low unemployment rates have prompted the company to hire entry-level international employees. These employees are hardworking and dedicated to their jobs, but their inability to understand and speak English sometimes causes problems on the production floor. Whenever technicians need to explain something to an international worker—whether about job training, production changes, or work to be completed when the line is down—they must find a translator. When no translator is available, workers tend to indicate that they understand but often do something entirely different, demonstrating that their understanding was incomplete. As they gain experience on the job, international workers tend to learn to understand job instruction signals. However, they still have problems reporting production and quality problems. Sometimes they can report the problems vaguely by calling and pointing, but often they lack sufficient proficiency in English to be able to report anything. They also have difficulty in helping with the training of new coworkers.

Because of these problems, the company asked the ESP Program in the Division of English as an International Language at the University of Illinois at Urbana-Champaign for help. We began the work by assigning MA-TESL students in an ESP training class the tasks of carrying out a needs analysis for six production lines and offering nonnative-English-speaking employees a free 10-hour English course. An external evaluation of the course was overwhelmingly positive, and the company consequently decided to offer a full 44-hour course based on the following performance goals:

1. read, recite, and demonstrate understanding of the quality policies and process of the company as presented in the Quality Skill Block (a training manual produced by the company)

2. read, comprehend, and recite the ISO quality statement and practices of the company that relate to each worker's particular job

3. demonstrate comprehension of the safety measures and practices of the production lines as presented in the Safety Skill Blocks (training manuals produced by the company) and safety procedures specific to each worker's job to ensure minimum accident rates

4. demonstrate the ability to read safety signs written in English

5. enable communication between the international employees and the native-English-speaking supervisors in order to enhance the relationship between employees and supervisors

6. improve communication among all production employees in the company and thus enhance the acceptance of international employees among native-English-speaking employees

7. enable international employees to comprehend basic training language in order to decrease training time and increase comprehension of simple directives

The Workers

The company selected 15 workers for the first class based on their work ability, their work records, and their need for further training in English. The workers' ability to read and write in their native language varied from none to a 12th-grade level; the average reading level was 3rd or 4th grade. Their ability to speak English ranged from no ability to minimal ability to understand simple instructions and communicate basic needs. No worker could meet the company goals stated above before taking the class. Two of the 15 workers were registered in an ESL class at a community college that, the workers claimed, emphasized general English and grammar and offered little help in improving communication at work. The other 13 workers neither attended nor planned to attend any English class. Most either worked many hours of overtime or held two jobs, which allowed little time for other activities. They spoke their native language at home. The low average reading level in the native language seemed to fuel a lack of interest in, a distrust of, and a fear of any type of schooling. Long working hours and little time to practice English meant that the class required a combination of creative approaches before it could begin to interest the learners to any great degree.

The low educational background of workers, however, did not indicate a lack of intelligence. On the contrary, many learned very quickly. They lacked formal education merely because in the past school had been too far from their homes or they had had to drop out of school in order to work to provide food for their families. Though the workers were intelligent, the low literacy levels and a 44-hour time limit meant that we had to illustrate concepts with photos, drawings, and realia and use as many hands-on and total physical response activities as possible to make learning successful.

◈ DESCRIPTION

The initial needs analysis carried out with company management, various line technicians, translators, and key workers resulted in a proposal for a 44-hour program containing six modules and related competencies (see Figure 1). Module 1, Getting Started, introduced language common to most companies plus this company's quality policy. Modules 2–5 dealt with specific language used in the company, and Module 6 involved communication coworkers, including supervisors. Key activities associated with the modules are described below.

Work Processes (Module 2)

Module 2 involved language that technicians use in training new workers and assigning job tasks. Once trained, these workers in turn help train other new workers, including native speakers of English. The company has 36 production lines, no two of which are identical. All 15 participants came from different lines and worked on other lines only when their own line was down.

Line Diagrams

To help workers understand their line, we taught them how to ask line technicians for names of machines and tools. They then drew simple diagrams of machines on their line and took the diagrams to their technicians to ask for the names of machines

Module 1: Getting Started

1. Greet coworkers, your boss, and other work acquaintances

2. Introduce yourself and ask someone's name

3. Use common courtesy words

4. Count to 100

5. Understand instructions to go to a particular line or a particular area

6. Ask the name of something

7. Clarify instructions and names by repeating essential parts, identifying unclear parts, asking someone to repeat or show them, and requesting someone to slow down or speak louder

8. Verbally communicate understanding or lack of understanding

9. Report a personal need (e.g., take a break, drink water, or use the restroom)

10. Read, recite, and demonstrate understanding of the quality policy and process of the company, extracted from the Quality Skill Block

Module 2: Work Processes

1. State names and describe the functions of machines, tools, and parts used on your job

2. Describe orally what you do to perform your job

3. Give others simple instructions to perform tasks or procedures related to a job

4. Ask/answer questions about locations in the workplace (such as tools, safety equipment, cleaning equipment, and people)

5. Request more boxes, dividers, strapping, water, a fire can, or other supplies

6. Identify potential jam points and explain how to unjam a machine

7. Explain what you should do when the line is down

Module 3: Work Orders, Shipping Labels, and Shipping Sheets

1. Name colors used for production

2. Name different types of packaging

3. Read work orders written in English to obtain the following information:

    a. the production line on which the product is to be made

    b. the number of units to be completed

    c. the product structure description

    d. any special instructions/requirements from the customer

    e. the color of the finished product

    f. whether or not the product requires special treatment

    g. the label type

    h. the number of boxes in a unit or on a pallet

    i. the position of the packed product (up or down)

    j. whether the finished product requires dividers, stretch wrap, bands, or pallets

4. Read shipping labels and tell how many are left

Continued on page 151

Module 4: Quality Measures Related to Work

1. Read, comprehend, and recite the ISO quality statement

2. Explain quality checks related to your job tasks or procedures

3. Report quality problems for production components and report problems with these components

Module 5: Safety Measures of Work Processes

1. Respond appropriately to oral warnings about potentially dangerous situations

2. Warn a coworker of imminent danger (e.g., with forklifts and machines)

3. Read and explain safety signs

4. Name personal protective equipment and ask about its location

5. Identify machine guards and safety devices on a line and state whether or not they are in place

6. Identify potentially dangerous parts of the machine on their line (pinch points, parts that start or stop automatically, high heat, sharp edges) and describe what to do when near these parts

7. Identify emergency stops and explain their impact on the machine

8. Report bad wires, faulty machine guards and safety devices, broken electrical or mechanical components, a fire, an accident

9. Ask where a fire extinguisher is

10. Report missing or damaged labels on chemical containers

11. Report spills and take action to clean them up

12. Tell the team leader, the safety director, or a human resource representative if blood or other body fluids from an accident have touched your skin

Module 6: Communicating With Supervisors and Other Employees

1. Order food in the company cafeteria

2. Ask about coworkers' families and activities

3. Participate in simple small talk (e.g., about personal background, family, the weather, weekend activities, favorite activities, recent activities or events)

FIGURE 1. Program Modules and Related Competencies

they did not know. Next, we discussed these machines in class and explained their functions, for example,

> This is the conveyor. The conveyor moves [the product].
> This is the case packer. The case packer puts [the product] in boxes.
> This is the ink machine. The ink machine puts the date on boxes.

Workers then taught each other the names and functions of machines on their assigned lines.

Tools and Materials

To introduce tools and materials, we made a symbolic drawing for every type of material or tool used on the line, pasted these drawings around the edge of a piece of paper, and wrote names of materials and tools in the middle. As each item was presented to the class, workers drew lines to connect the names to the respective tools and materials. Next, workers in groups of four received a packet of cards with individual illustrations of all tools and materials. As workers turned each card over, they practiced naming the items. All group members were responsible for helping each other learn. The next day they practiced with the same items, this time requesting materials or asking about the location of an item, for example,

> I need more cardboard.
>
> I need more labels.
>
> We're out of strapping.
>
> Where is the fire extinguisher?

Some workers simplified these phrases to *more cardboard* or *more strapping*. Others learned the entire phrase. All learned to communicate needs they could not previously express to other line workers or technicians or via the auditory paging system.

Job Tasks

After workers discussed the job tasks performed on their lines, we took digital photos and combined these with symbolic drawings to illustrate each job task. Workers used the resulting list of 78 tasks and illustrations to select the tasks they performed on their own lines. Then they cut out the illustrations and taped them on white paper to create a set of job instructions for their individual lines. Working in pairs, they used the page(s) of illustrations to practice giving instructions to other workers in the class, for example,

> Unjam the box machine.
>
> Stack the boxes six high, four wide, and three deep.
>
> Strap the unit.
>
> Put a shipping label and a shipping sheet on the unit.
>
> Tell the technician if there is a problem.

Again using illustrations and line diagrams, they practiced training each other in how to unjam machines on their line.

Illustrations used in combination with written names and instructions enabled all workers to practice training other workers in machine names and functions and in instructions for common job tasks. Those who had at least a third-grade reading level began reading in English, using associated illustrations to improve reading comprehension. Workers improved both listening and speaking related to work on their own lines while training other workers in the class. Simultaneously, as they listened to other workers comment about their own lines, each worker in the class received cross-training on new lines to which they might be assigned when their own lines were down.

Work Orders, Shipping Labels, and Shipping Sheets (Module 3)

We used drawings in all aspects of Module 3, even when checking the reading comprehension of work orders. The use of illustrations guaranteed that, for quality purposes, workers understood what they read instead of merely matching words to find an answer. For example, rather than ask whether a unit required plastic wrap, we had workers look at a work order and then at an answer sheet with an illustration of plastic wrap followed by the words *yes* and *no*. Workers needed only to circle the correct answer or occasionally fill in a number, as when responding to the question *Number of units?*

Again workers often worked in pairs. After learning to read work orders, for example, they received a completed answer sheet and transferred the information on it to a blank work order. They also compared work orders, shipping labels, and shipping sheets, first to see where information on shipping labels originated on the work order and later to find mistakes on shipping labels and shipping sheets that did not correspond to work orders.

Quality Measures Related to Work Processes (Module 4)

The company's emphasis on quality and ISO 9002 certification meant that workers had to be able to identify and report quality problems. As in Modules 2 and 3, we worked with illustrations and realia to teach and practice the names of specific quality problems. Pair work and group games provided intensive practice in naming problems and identifying problems that were named. We worked on reporting problems not only with products produced by the factory but also with labels, packing materials, and packaging, always emphasizing the necessity of identifying problems immediately. To avoid the reworking of finished products to meet customer requirements, everyone was responsible for reporting quality problems.

Safety Measures of Work Processes (Module 5)

Good safety procedures create a win-win situation in which workers are not hurt and the company's worker compensation costs do not increase. When we began Module 5, workers were unable to read most company safety signs. They also had difficulty in reporting machine problems and accidents, explaining where they hurt, and warning others of potential danger.

To teach the reading of safety signs, which typically consist of a series of words with no pictorial symbol, we created colorful symbols to illustrate the meaning of 63 signs appearing around the plant and taped them on the wall on 2 separate days in four groups of eight signs. Then we gave each worker a list of the words corresponding to the signs on the wall that day and explained the meaning of each sign, using a transparency to present the symbolic illustration as we read the words with the workers. Next, the workers walked around the room in small groups, putting the number of the illustration next to the corresponding words on their list and competing to see which group could finish first. On another day, we gave workers several sheets of paper with two columns of symbolic illustrations and a column of typed signs in the middle. Workers drew lines to connect each illustration with the words expressing its meaning.

The reporting of aches, pains, and accidents was treated similarly. After an introduction to terminology, workers in groups of three received a set of cards, each displaying an illustration and a label, such as *headache, backache, stomachache, cut, fall, burn,* and *eye injury.* Workers then took turns drawing cards from the stack and pantomiming the illustrated problems while other workers in the group named the problems and reported them. On another day, the workers role-played accidents, warning people of danger, calling for help, reporting the accident to a supervisor, and going to the doctor. We had the workers practice each scenario in large and small groups to give them as many chances to speak as possible. Sometimes we videotaped their role plays and showed them to the class.

To assist workers in communicating in English about personal protective equipment and potentially dangerous parts of machines on their lines, we again used realia and drawings, and referred to the line diagrams used at the beginning of the program. Workers marked potential dangers in red and identified the machines and potential danger. They also practiced reporting problems and giving instructions to avoid accidents, for example,

> The trimmer can cut. Use special gloves when you unjam the trimmer.
>
> Don't stand on the conveyor. You can fall.
>
> The wire is bad.

Expressions practiced in this module reviewed the machine names, tools and materials, and job instructions learned in Module 2.

Communication With Supervisors and Other Employees (Module 6)

Module 6 incorporated some aspects of general English as workers practiced talking about their families and activities. To teach workers how to order food in the cafeteria, however, we customized the activities by going to the cafeteria, requesting a list of foods generally sold there, creating a series of illustrations of each food, and reproducing them on cards. When introducing new material, we gave them handouts with illustrations and related names or phrases, but for practice we always used illustrations and realia. In this way, workers who could read well benefited by having both written words and illustrations, and those who could not read always had illustrations as an aid to review with their children and friends.

One other distinctive aspect of the last module involved daily and weekly news. After the first week, we spent some time in each class discussing news with the workers. Their favorite topics tended to be job related, such as "good production," "bad day with the line down for 2 hours," and other work-related matters of interest. We wrote up the news and brought it to class the next day, making sure to include at least one comment from every worker. At first we read the news aloud with the class. Later, workers began reading their own printed news aloud, with varying degrees of pronunciation help from us. The entire class understood that English is phonetically different from their native language, which made reading together and accepting help much easier.

◈ DISTINGUISHING FEATURES

Several characteristics of our program exemplify successful ESP programs in the workplace. First, we, the ESP staff, clearly understood the needs and limitations of the clients and responded in reasonable ways. Second, we developed appropriate materials and delivered effective instruction. In addition, our efforts generated positive, observable results.

Understanding Clients and Responding to Their Needs

Teaching entry-level workers in manufacturing, who have an average third-grade reading level, a distrust of schooling, a fear of embarrassment when speaking English, and little motivation to learn or use any English that is not directly related to improving wages, offers a special challenge to ESP teachers. To meet this challenge, the program combined oral English with realia and labeled illustrations, employed pair and small-group practice, and made learning enjoyable through a rich variety of useful and interesting activities.

As a result, workers became highly motivated by the program and kept requesting that classes be repeated so that others could enjoy them, too. Learners felt not only that they had learned English but also that the production-based materials had improved their understanding of their work. They were happy that the class allowed them to individualize much of the language learned to their particular lines, and they appreciated the opportunity to identify their own learning needs and have them addressed in the program along with company-specified material.

Developing Appropriate Materials and Delivering Effective Instructions

In addition to the various instructional activities described so far, the ESP program developed a multimedia program on CD-ROM for training entry-level workers—both native and nonnative speakers of English—to understand oral instructions for line production, safety, and quality and to read safety signs. Learning this language is crucial for high-quality, safe, and efficient work. The program was first developed in English and Spanish; more languages may be added in the future.

The program contains the following three modules:

1. reading comprehension of 63 safety signs around the plant, listening comprehension of oral warnings of potential dangers on the line, and safety precautions to be followed to avoid accidents

2. listening comprehension of names of materials and oral instructions to perform specific job tasks on each line

3. reading and listening comprehension of the company's quality statement; listening comprehension of potential quality problems with the product, labels, and packing

All new workers complete identical material in Modules 1 and 3. Module 2 contains a 36-item production-line menu so that workers may select the line to which they have been assigned.

Each module contains three sections. In the first, a teaching section, workers see photos, drawings, diagrams, a work order, or a shipping label on the screen. They click the mouse on an illustration or phrase to hear and see the corresponding name of the item, quality problem, or job instruction in English or to hear the pronunciation of a word or phrase in the work order and see an illustration of its meaning. If they need the meaning in their native language, they click an interpreter icon to hear the word or phrase in their own language and hear it again in English.

In Section 2, a practice section, workers hear groups of terminology, phrases, and instructions and perform a task, such as clicking the mouse on a photo, drawing, or section of a diagram, to show that they understand what they have heard. They then hear a word or phrase and see it written on the screen. Again, clicking on a translator icon allows workers to hear the phrase in their native language. The quality section asks workers to play against a clock, identifying quality problems as quickly as possible. The safety section asks them to drag safety signs on top of their corresponding illustrations and symbols.

The third section is a test in which workers hear terminology, phrases, and instructions in random order and perform a task similar to that in the practice section. In contrast to the practice section, workers hear the phrases only one time and no longer have access to the translator. The program keeps records of progress and problems, which may be accessed by the training director. When the score is high, workers receive a bonus coupon (e.g., for a soft drink or cup of coffee in the cafeteria, or a company T-shirt), an award certificate, or a message that may be taken to the training director or line technician to show that the worker is ready for a reward and for further evaluation of the ability to understand the language on the job.

Generating Positive, Observable Results

The program resulted in improvements in workers' individual learning and in their job performance, and the company's return on its investment in the program was high.

Individual Learning

All workers, even those with no or minimal literacy in their native language, showed improvement in final tests of learning (see Table 1). These tests, which involved both ESP teachers and company trainers, continued the use of illustrations to check individual proficiency in reading work orders and safety signs and in giving job and safety instructions, reporting problems, and ordering food in the cafeteria.

Change in Job Performance

Nine technicians evaluated workers' change in performance on the job. Question-naire results (see Table 2) plus interviews with line technicians and other immediate

TABLE 1. AVERAGE SCORES ON INITIAL EVALUATION AND POSTTEST FOR COMPETENCIES IN MODULES 2–5 (%)

Competency	Initial Evaluation	Posttest
Read work orders	40	71[a]
Read safety signs	7	79[b]
Talk about machines, danger points, job processes	9	86
State the company's quality principle, report quality and machine problems, request materials	3	88

[a]Two workers scored 100%.
[b]Seven workers scored 93% or higher.

TABLE 2. TECHNICIANS' AVERAGE RESPONSES TO QUESTIONS ON
WORKERS' JOB PERFORMANCE (%)

Question	Answer	
	Before Class Began	Now
How well was this worker able to understand your instructions?	58	89
How much did this worker speak English to report problems, ask for information, and ask questions?	60	82
How much has production increased?	73	
When lines are down, how much has worker initiative to clean, etc. improved?	84	
How much has motivation increased?	82	

supervisors showed definite changes in job performance, with workers asking more questions, translating for other workers, reporting more problems, and starting to train new workers who were native speakers of English. Only one worker, a woman who could not read or write in her native language and who remained very shy in using both English and Spanish, demonstrated little change in job performance. The technicians offered comments about six of the workers:

When someone new comes on the line, he pitches right in to translate and train him/her [an unanticipated benefit]. He showed a lot of improvement in both speaking and understanding. Now if he doesn't understand, he asks questions.

Since the class began, she now always says good morning, converses more, and will holler and tell me if something is wrong. She is becoming the translator for other workers.

He now asks more questions and translates for other workers.

If [I talk too fast and] he doesn't understand, he now says "slow down." He really improved!

I have seen an improvement. He always tried but sometimes ended up in sign language. He now asks more questions about quality, for example, "Is this okay?" He also started asking a lot of questions about work orders. He always helped with new people in both languages and now does this more efficiently. He has helped increase production in the sense that things run smoother; there are smoother transitions with breaks and lunches and faster instructions to others to go to the heat tunnel, cut out labels, etc. The biggest change is in initiative. He knows more about what's going on. Now he will go to the toolbox and start the changeover on the box machine. He orders boxes and gets dividers.

Before if he didn't understand, he wouldn't say so. Now he lets me know. He has started asking a lot more about machines. Before, he learned a lot by watching. Now he asks questions. He has asked a lot more about the machines. I really like this. The class helped.

Return on Investment

The human resources director turned in an analysis of the company's return on its investment in the program (see Table 3) with a comment that benefits had far exceeded the cost. Another possible source of return on investment that remains to be investigated is reduction in turnover. Human resource directors in Illinois claim that the cost of hiring one entry-level worker can be as much as $2,700 because of drug testing, interview time, and training. When turnover is reduced, this money is saved. (See Phillips, 1997, for more information on ways to calculate return on investment.)

◈ PRACTICAL IDEAS

Various elements that helped make this program successful were (a) the use of digital photos and drawings, (b) worker participation in needs analyses and instruction, (c) the introduction of something new in every class, (d) frequent review, and (e) a combination of ESP and content-based instruction.

Use Digital Photos and Drawings

A digital camera is ideal for creating illustrations for classes and a teaching program on CD-ROM. With such a camera, one can see, reduce, enlarge, and edit photos immediately on the computer without incurring the extra costs of film and development. Because companies have proprietary restrictions, a company manager such as the director of human resources or a plant or production manager should review the photos on a computer screen and approve them before they are taken out of the plant.

Enlarged photos and transparencies made from these photos work well for presentations to the entire class. Medium-sized photos work well for cards for small-group work and games. Reductions as small as $1\frac{1}{2}$ by $1\frac{1}{2}$ inches work well for handouts. Teachers should remember to save originals and always use them for photocopies, as a photocopy of a digital photo is clear whereas a photocopy of a photocopy is not.

Workers like to take handouts or booklets home, but they do not always remember to bring them to class. Teachers who plan to use a handout for more than a day—for example, workers' individualized job instruction sheets—might make two copies of the original and keep the original close at hand in case a worker accidentally takes home the training room copy.

TABLE 3. RETURN ON INVESTMENT IN THE ESP PROGRAM OVER 3 MONTHS

Area	Improvement (%)	Savings ($)
Accident reduction	1.5	585
Downtime reduction	2.0	1,376
Safety violation reduction	1.0	?
Communication barrier reduction	60.0[a]	1,500[a]

[a]Estimate based on technicians' evaluations.

In creating drawings to use in class, artistic ability is helpful but not necessary. Drawings should be symbolic and as simple as possible. The drawings shown in Figure 2 are typical of those used in this program. Note the small, somewhat elongated head in stick figures representing adults and the use of symbols for verbs: an eye means *check*, two parallel arrows indicate movement such as *push* or *lower*, a person with a big hand held out means *Please help me (give me a hand)*, and a flame next to a glove indicates a fire-resistant glove. Combining symbols with digital photos can help make a meaning clear. For help with drawing or information on appealing to multiple intelligences in teaching, see Shapiro (1994), Wright (1984), and Lazear (1991).

Elicit Worker Participation

Workers should participate in defining what they need to learn and in teaching others. Adults need some control over what they will learn. In many international cultures, sharing and helping others are more important for survival than competition is. Consequently, pair and small-group work tend to be very effective—workers learn more as they teach each other.

Teach New Material in Each Class

Workers need to feel that they will learn something new in every class. Overtime, two jobs, responsibilities for extended families, illness, and exhaustion compete with classes in a socioeconomic culture that has traditionally deemphasized education. Therefore, it is very important to make workers feel that they are not wasting their time. In every class, they should learn something new. Every minute of the class needs to be useful.

Review Frequently

Workers like the ones in our program frequently do not have a tradition of taking material home and studying it. Nor do they have much time to study. Though we

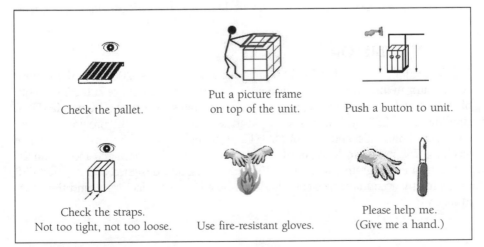

FIGURE 2. Use of Drawings to Explain Instructions

always gave the workers something to take home so that they could review it individually or with their children or friends, we gave no homework assignments except for a few instructions, such as "Bring a diagram of the machines on your line" (they could also create these diagrams during the class) and "Ask your technician for the names of machines on the line." The most frequent assignment was to ask or say something in English to their technician or to other people in the company. To replace study at home, we offered daily review in class. We reviewed job instructions for their lines, for example, on at least 5 separate days, giving workers the opportunity to practice speaking English by cross-training almost everyone in the class.

Employ Content-Based Instruction Along With ESP

Workers who do not understand English frequently have only a vague understanding of various aspects of their work. Consequently, the ESP program involved job training so that workers learned more about what their jobs entailed and what the company expected of them even as they learned English. This combination of ESP and content-based instruction for individual production lines, and the introduction of cross-training on other lines as workers practiced training each other, proved interesting and effective, helping the workers and the company. (For other ideas about ways to talk with businesses, discover content needs, and involve workers in decision making, see Auerbach, 1992; Chisman, 1992; Johnston & Packer, 1987; Rothwell & Brandenburg, 1990; Thomas, 1991.)

◈ CONCLUSION

Many ESL/ESP teachers despair when faced with the type of workers described in this case study. Program dropout rates are often high, workers' interest in learning general survival English is low, and companies often have little willingness to pay for ESP programs. Yet as this case study has shown, customizing the program to meet the company's production needs and the workers' learning needs can overcome all of these obstacles. Companies increase their bottom line, and workers beg to be allowed into the class. The company, its workers, and the ESP providers all benefit.

◈ CONTRIBUTOR

Judith Gordon, chair of TESOL's ESP Interest Section in 1999–2000, is ESP program coordinator in the Division of English as an International Language at the University of Illinois at Urbana-Champaign, in the United States, where she trains MA-TESL candidates in ESP and works with businesses to set up ESP programs, create customized materials, and mentor MA-TESL teachers. She has 23 years of experience teaching ESL in Puerto Rico and the United States. She has designed more than 20 ESL computer programs, written multiple ESL texts, and designed and directed ESP programs for accountants, science students, business, manufacturing, and the green industry.

CHAPTER 10

An ESP Program for Union Members in 25 Factories

Paula Garcia

◈ INTRODUCTION

Many workplace ESP programs exist in large corporations with overseas connections. These programs seek to improve communication within the global economy and tend to train well-educated employees on targeted topics for a short time. Much less common are workplace education programs designed for blue-collar workers. One program, the Worker Education Program (WEP), existed in the Chicago, Illinois, area of the United States from 1992 to 1997. It was created for the purpose of improving workers' English language skills. By the time funding ended, the program had offered 18 different courses to more than 1,300 workers in 25 companies. This chapter describes the important elements of this historic program from inception to completion, covering issues such as student recruitment, support services, teacher selection and training, pedagogical philosophy, program evaluation, benefits, and outcomes.

◈ CONTEXT

The WEP came into existence as a possible answer to the desire of a labor union (the Amalgamated Clothing and Textile Workers' Union, now called UNITE) to improve communication with and participation from its nonnative-English-speaking members. The objective was to provide workplace education programs for union members in the Chicago area. Fulfilling the objective required a two-tiered system for delivering instruction: classes in individual factories and classes at the union hall. On-site classes were held at partner companies that had the facilities for instruction, such as a well-lit cafeteria, a blackboard, and other equipment. For union members working at companies that lacked instructional facilities, classes were held in union hall meeting rooms. The WEP was concerned about providing instructional services to all union members, and the variety of class sites served this purpose.

The Companies and the Union

The Chicago area presents a special challenge for small companies because much of the workforce in nonskilled jobs comprises immigrants and nonnative English speakers. The changing nature of nonskilled labor requires strong communication skills in oral and written language (Carnevale, Gainer, & Meltzer, 1988). The introduction of new production techniques such as statistical quality control and computerized machinery has unveiled serious communication problems in a number of manufacturing companies. WEP partner companies cited oral communication and literacy skills as critical factors in their transition to alternative production models. These changes in manufacturing alerted the union to the need for improved basic skills. The union realized that unless it could increase the educational level of its members, productivity would decrease, which would ultimately lead to a weakened union. With help from educators at Northeastern Illinois University and the Chicago Teachers' Center, the labor union acquired a grant from the U.S. Department of Education's National Workplace Literacy Program and launched the WEP.

The provision of educational programs for union members is not new in organized labor. The Amalgamated Clothing and Textile Workers' Union first began providing English classes for its membership in the early 1900s. Union meeting notes from this early era were written in seven languages, indicating that managing a multilingual group of people is a union tradition. To start the WEP in 1992, business agents (i.e., the union's liaisons with companies) collaborated with shop stewards (i.e., factory workers who manage the union locals) to approach company management and employees about setting up an educational program that incorporated everyone's needs. Many employers at this time were grappling with changing technology and new production paradigms, and welcomed an educational program funded and organized by the union. Employees were interested in increasing their basic skills for personal and job-related reasons.

The Workforce

The ethnic mix of the U.S. labor force today differs considerably from that of the early 20th century. Instead of masses of Italian-, German-, and Yiddish-speaking workers, the non-English-speaking union membership now comprises mainly speakers of Spanish, Chinese, and Polish. The WEP attracted student workers from Mexico, China, Guatemala, Russia, Italy, Poland, Ukraine, Colombia, Haiti, Cuba, and other countries (Boyter-Escalona, 1998).

Most important for any workplace education program is to include all workers, no matter what their educational needs are (Sarmiento & Kay, 1990). At its inception, the WEP decided to address the needs of all workers in its educational services. Native English speakers were included as well through courses that prepared workers for the General Educational Development (GED) test or through computer literacy classes, both of which became part of the repertoire of WEP classes. In addition, partner companies and the union often asked the WEP to offer targeted minicourses on topics such as team building, conflict resolution, and health and safety. The particular makeup of each company determined which courses were offered.

◈ DESCRIPTION

Using a partnership model, the WEP built a solid foundation on which to develop and expand. After advisory boards consisting of company personnel, union staff, and educational staff were formed, program coordinators (of which I was one) could perform task analyses, recruit students, hire teachers, and put the educational programs in place.

Recruiting Partner Companies

Before the WEP could recruit students, it had to recruit partner companies that would be willing to offer on-site workplace literacy classes. Not all companies organized by the labor union were willing or able to do so. Some were quite small (i.e., fewer than 100 employees) and simply did not have the resources to offer on-site classes. For these companies, day-to-day survival was a struggle, and they could not contribute the time and effort to participate. Because of cutbacks and lack of resources, several companies did not consider learning as a priority. Ironically, these companies would have benefited even more than the others from an educational program.

WEP staff had to develop a sense of whether a company was ready to implement a workplace education program. One company initially refused the offer of free on-site ESL and GED classes for workers because of decreases in profits and production. This company had recently laid off several employees and did not think the time was appropriate. Two months later, business improved, and the company recalled laid off workers and hired additional ones. This change created an opportune time for WEP classes because the expanding company suddenly needed better skilled workers and was experiencing difficulty in finding them.

Another hurdle for the WEP was dealing with the traditionally adversarial relationship between the union and employers. On rare occasions, the WEP director, the teachers, and the other program coordinator and I found ourselves walking a fine line between the conflicting agendas of the two entities. We had to remind the teachers and ourselves that we should not choose sides in such conflicts; all WEP staff had to remain politically neutral. However, most conflicts were resolved because the union and the company shared a primary interest in production and profit increases, which were possible benefits of the educational program. What was good for the company was usually also good for the union, and having an educated workforce was in the company's and the union's best interest (Sarmiento & Kay, 1990).

An important first step in setting up educational programs at company and union sites was to construct an advisory board at each company comprising company personnel, union staff, and educational staff. Advisory board members from partner companies included chief executive officers, safety managers, production managers, human resources personnel, and line workers. Advisory board members from the union were the educational director, business agents, and shop stewards. The WEP staff included the director, program coordinators, and teachers. These advisory boards proved to be instrumental in developing true partnerships among unions, companies, workers, and educational personnel.

Recruiting Students

Student recruitment was a constant endeavor for us as program coordinators. Union business agents, shop stewards, and teachers contributed to the various recruitment activities. With the cooperation of company management, we, along with business agents, organized open-house events, which often took place in the company cafeteria. At these events, workers could meet the teachers and us face-to-face, ask questions about the educational courses, and become more comfortable with program staff. Other recruitment strategies included passing out flyers and making brief presentations during lunch breaks.

These recruitment strategies were useful in getting the program off the ground, but once it was established, constant visibility became an equally effective way to attract new students. The WEP held graduation parties at the end of 12-week courses and minicourses, which attracted attention to the educational program. Also, WEP students often received recognition from the company in the form of bonuses and job promotions, which made other workers want to take advantage of classes, too.

One obstacle to recruitment was a general negative impression of learning in the minds of participants. Most WEP students had had less than 8 years of schooling in the United States or their native country, and many reported having unhappy memories of school. In several partner companies, workers regarded attending WEP classes as detrimental to worker solidarity. They thought of WEP students as workers who were trying to surpass the rest of the workforce and climb into middle management positions. Workers who were sensitive to this perception tended to stay away from classes even though they might have wanted to participate. The WEP attempted to deflect these perceptions by involving union staff and shop stewards, who spoke personally with students affected by these attitudes.

Another obstacle was workers' low self-esteem, which is a common problem in workplace programs (Soifer et al., 1990). Many workers perceived themselves as too old or ignorant to learn. After having been away from a learning environment for so many years, many workers felt that beginning to think critically again would be too difficult for them. WEP staff constantly reassured these workers and pointed to participating workers as examples of students who did not let their low self-esteem prevent them from learning. In fact, participation in the program often had the effect of increasing workers' self-esteem (Boyter-Escalona, 1998).

Another deterrent to participation was that partner companies were unable or unwilling to grant workers paid release time for the educational program. This meant that the WEP had to offer classes on workers' own time, before or after work. The WEP's sponsor recognized the hardships that might be placed on workers who wanted to attend classes outside of work hours and supplied funding for support services, such as transportation vouchers and child care stipends, to overcome such problems.

Selecting and Educating Teachers

Finding ESL and GED teachers is a challenge for any educational program, but especially for workplace programs because so few teachers have workplace experience (Soifer et al., 1990). Because of the nature of workplace programs, teachers must be flexible and innovative. Workplace classes are often conducted at odd hours of the day and night in factory cafeterias and meeting rooms, which may not be conducive to traditional classroom learning. When hiring, the other WEP coordinator and I informed prospective teachers of the nature of on-site teaching during the interview. This practice curtailed any possible misunderstandings and misconceptions of what teaching in a factory entailed.

One of the challenges of managing a union-based educational program is finding teachers who are sensitive to and knowledgeable about union issues and concerns. To tackle this problem, the WEP offered in-service sessions on union topics and occasionally invited union personnel to teachers' meetings. Not only did teachers have to be willing to learn about union issues, but they also had to be interested in learning about their students' jobs. Because they were teaching ESL and literacy skills for the workplace, they had to have a comprehensive understanding of the various departments and positions on the factory floor. Before they began teaching at a work site, teachers were led through a tour of the facility. In addition, the curriculum guides designed for individual companies gave teachers information on specific jobs.

Workplace teachers needed to be sympathetic toward students who were tired and slow to learn after a day's work. Rather than begin class with lessons that were overly challenging for students, teachers usually began with reviews of previous lessons and interactive conversational activities that eased students into more mentally challenging tasks. During class, teachers were patient with students' response times and offered one-on-one guidance.

Supporting Teachers

WEP teachers were supported by monthly teachers' meetings, which the other program coordinator and I organized and chaired. Because teachers worked at a variety of partner companies, they were separated from each other and from WEP program staff. The weekly meetings alleviated teachers' feelings of isolation and were a forum for sharing lesson plan ideas and the ups and downs of teaching in the workplace. The monthly meetings incorporated in-service training sessions on topics such as integrating workplace forms and shop floor realia in classroom instruction, and teaching multilevel classes effectively (Boyter-Escalona, 1998). Teachers received compensation for attending any meeting, including advisory board meetings, and for preparation time (1 hour for every 3 hours of instruction).

Because of the challenging nature of teaching in the workplace and the lack of experienced workplace teachers, the WEP hired a cadre of teacher aides. Aides helped head teachers deal with multilevel classes and were trained to become head teachers by observing and following the direction of the teachers. This practice provided an experienced pool of teachers for the WEP to select from as new workplace sites were organized.

The WEP director and coordinators were very concerned with the quality of teaching in the workplace and union hall classes. We observed teachers regularly and gave detailed feedback on delivery and organization of lessons. Teachers handed in written lesson plans, which we checked for workplace content and learning activities. Although the program supplied books and materials for teachers and students, we encouraged teachers to develop their own instructional materials because commercial materials were not always appropriate for tailored workplace classes. The sharing of lesson plans and work-related materials eased some of the struggles associated with creating customized materials.

Evaluating the WEP

The WEP evaluated students' progress by means of several standardized testing measures that had been approved by the U.S. Department of Education. The program director used scores from these tests to report the students' learning gains and evaluate the program's effectiveness. These reports indicated steady increases in students' basic skills; unfortunately, none of the standardized tests assessed work-related skills. WEP teachers pointed out that the tests did not validly assess learners' gains because they did not focus on workplace vocabulary, reading, or writing, which were the main components of the classes.

Concerned about the mismatch between what was being taught and what was being tested, the other WEP coordinator and I, with input from the teachers, created the General Work-Based Assessment (GWA; Martin & Garcia, 1997). This test measured English language ability in a manufacturing setting through an oral interview, a listening comprehension section, a reading comprehension section, and a short essay. All tasks in the GWA were work related. We piloted the test at several WEP locations and, on review by the U.S. Department of Education's Workplace Literacy Program, it was accepted as a measure of WEP students' basic skills. In fact, such a great need existed for workplace-specific testing that the Illinois secretary of state, whose office also funds workplace literacy programs, accepted the GWA for use in its programs.

Psychometric testing was only one of the methods used to evaluate program effectiveness. WEP staff also collected qualitative information from workers and floor managers on the extent to which communication improved in the factories as a result of WEP classes. Company managers answered questionnaires regarding improved employee performance, increased safety awareness, decreases in accidents and grievances, reduced scrap and waste, worker promotions, and improvement of communication and math skills used on the job. Union business agents and officials also answered questions on the usefulness of the educational program. They reported that more workers were participating in union meetings and contract negotiations and taking advantage of union services as results of the WEP (Boyter-Escalona, 1998). Overall, the companies and the union reported substantial gains resulting from increased productivity and decreased accidents and mistakes. These bottom-line gains proved to partner companies and the union that educational programs were valuable additions to the workplace.

◈ DISTINGUISHING FEATURES

The WEP stood out in its field in a number of ways. Besides the fact that the program was conceived of and implemented by a labor union, other distinguishing features were its pedagogical philosophy, the method of conducting needs assessments and task analyses, the multifaceted curriculum guides, innovative classroom techniques, the development of classroom materials focusing on a variety of workplace topics, and attention to workers' additional learning needs.

Pedagogical Philosophy

The WEP decided early on to adopt a holistic, student-centered approach to curriculum development and syllabus design. In workplace education, *student centered* means *worker centered,* and it is the optimal approach in a union context (Sarmiento & Kay, 1990). A worker-centered approach links education to workers' needs and interests, both on and off the job.

As mentioned previously, the changing nature of factory work gave rise to the need for the WEP. To be prepared for these changes, workers needed to improve critical thinking skills as a first step. Partner companies stated repeatedly that they wanted workers to have these skills because they facilitate the transfer from one method of manufacturing to another. Short-term training programs that focus exclusively on job skills do not effectively address critical thinking skills; they are usually no more than a quick fix (Sarmiento & Kay, 1990). A worker-centered approach integrates workers' needs and interests with the educational goals and objectives of the program.

Needs Assessment and Task Analysis

The first step in carrying out a worker-centered approach is to assess the needs of all parties involved in the program: employer, employee, and union (Sarmiento & Kay, 1990; Uvin, 1996). The WEP offered educational programs at a variety of partner companies involved in a variety of businesses, from manufacturing lighting fixtures, paint, ballet shoes, men's suits and shirts, and medical supplies to providing services such as industrial laundering and silk screening. This variety of workplace contexts required the WEP to study many types of jobs and systematically assess the needs of each plant. Program staff interviewed workers, floor managers, line supervisors, maintenance personnel, area supervisors, and others to find out what communication difficulties companies faced. These interviews were instrumental in acquiring an in-house perspective of attitudes toward learning. For example, at several plants, we were informed that some workers were afraid of company management's finding out about their low literacy skills. Once we had been warned, we could adopt a strategy for minimizing fears, such as repeatedly reassuring workers of absolute confidentiality.

Immediately following the needs assessment, we conducted task analyses to acquire a comprehensive understanding of how each company organized its factory floor. Task analyses involved factory tours, interviews of workers and employers, observations of workers on the job, and reviews of company forms and informational materials, such as training manuals and brochures. Teachers were as involved in this process as was possible.

From the task analysis, we determined the types of reading and speaking skills employees used and needed to use on the shop floor. We culled work-specific words and phrases for classroom use. Most importantly, we learned about the overall structure of each workplace so that we could relate to workers, via teacher instruction, how each worker fit into the company's overall organization, which company and union management had identified as an essential quality of a well-run workplace. Workers who fully understand how their performance on the job directly relates to other workers and to production are more likely to generate greater enthusiasm and effort for their part in the whole.

Another important part of task analyses was learning specialized vocabulary. Many companies used technical terms for machinery, parts, and processes. In some cases, workers had developed their own vocabulary words that conflicted with the company's words. The educational program resolved the resulting breakdowns in communication. For example, in one company, workers used the Spanish word *cuerno* (horn) to describe a certain part that reminded them of a miniature bull's horn. The company used an English word for this part on work order forms and assumed that workers knew it. In WEP ESL classes, the workers learned the English word for the part and learned to understand it on work order forms.

Additionally, educational programs must first alert all employees and union staff before conducting any needs assessment or task analysis. The WEP kept all parties informed through the union's business agents, union stewards, plant or floor managers, and advisory board committee members. If this step is neglected, educators may encounter animosity from the workforce, creating a difficult and undesirable situation for any educational program (McGroarty, 1993; Sarmiento & Kay, 1990). The WEP rarely faced opposition when conducting needs assessments or task analyses because it was affiliated primarily with the union.

Multifaceted Curriculum Guides

Using information collected from needs assessments and task analyses, the other program coordinator and I developed curriculum guides for each partner company as well as the *General Workplace Curriculum Guide* (*GWCG*; WEP, 1994), a general guide that could be used across programs. The *GWCG* was not customized for a particular company. It was developed for the union-based classes, which consisted of workers from various plants. It also served as a jumping-off point for individual curriculum guides, and it was a model shown to companies interested in beginning on-site workplace programs. Individualized company curriculum guides listed course goals, learning objectives, vocabulary items, and lesson plan ideas.

The *GWCG* was divided into thematic units: "Communication in the Work-place," "Health and Safety," "Quality Control," "Work Forms," "Company Rules," and "Vocabulary and Expressions Used on the Job." Each thematic unit listed pedagogical goals: language skills related to the theme, lesson ideas and activities, and materials. For example, one of the goals listed in the unit "Communication in the Workplace" was "to communicate problems at work to appropriate person, orally and in writing" (WEP, 1994, p. 19). Language skills related to this goal included present and past continuous verb tenses and the expression of needs and wants. Related lesson ideas included brainstorming common workplace problems, role playing, and completing strip stories. The materials column cited textbooks and gave page numbers for lessons on this topic and other materials, such as flash cards. The ideas for lesson plans in the *GWCG* was were an important lesson-planning resource for WEP teachers.

Innovative Classroom Techniques

The WEP's worker-centered approach involved participatory classroom techniques that transferred ownership of the classes to the workers themselves. As stated in the *GWCG* (WEP, 1994),

> The worker-centered, or participatory, approach links education to workers' social realities where they take an active role in their own learning. Teachers do not serve as problem solvers; rather, they are problem posers. The responsibility of looking for solutions belongs to the workers, which builds their capacity to solve problems and directs their future lives. Because this program was initiated by the union, and the union is the members, these classes belong to the members. This kind of ownership gives workers an active part in their own education, hones their decision-making skills, and builds their self-confidence, thereby enabling them to participate more fully in the workplace. (p. 6)

As problem posers, WEP teachers used Freirian techniques (Freire, 1973) to build students' capacities to identify causes of and possible solutions to communicative problems in the workplace. Teachers utilized role plays and dialogues that encouraged students to identify communication problems and gaps in their abilities. The goal was for students to feel comfortable discussing what they needed to learn so that teachers could design lessons that addressed those needs.

Because the majority of student workers in the WEP's ESL classes averaged 8 years of schooling or less, traditional grammar-translation and structural syllabus models were inappropriate. Students did not view language as comprising discrete categories of words; therefore, teaching the structure of, for example, the various verb tenses was not a successful technique. WEP teachers therefore became adept at applying communicative approaches, such as the notional-functional approach, content-based instruction, and task-based language teaching (see Krahnke, 1987).

One common classroom technique involved the use of realia from the factory floor to teach various grammatical functions. For example, one instructor used parts from the production line to teach prepositions. Students took turns giving brief instructions to one another, such as "place the back brace next to the connector" or "put the pins inside of the flange" (WEP, 1994, p. 55). Teachers used vocabulary items elicited from the task analyses in content-based lessons, for example, filling out order forms for parts or writing dialogues about machinery defects. Task-based activities such as "Spot the Difference" and information-gap exercises were customized for the workplace. The workplace served as an ideal setting for English language instruction in the context of work-related vocabulary and concepts.

Materials Development

Few work-related ESL textbooks exist for beginners. The WEP ordered multiple copies of just about every workplace literacy and ESL textbook available (see Figure 1), and teachers checked them out as needed. Teachers used these texts

Auerbach, E., & Wallerstein, N. (1987). *ESL for action: English for the workplace*. Reading, MA: Addison-Wesley.

Auerbach, E., & Wallerstein, N. (1987). *ESL for action: Problem posing at work*. Reading, MA: Addison-Wesley.

Brems, M. (1991). *Working in English: Beginning language skills for the world of work*. Chicago: Contemporary Books.

Gordon, J. (1991). *More than a job: Readings on work and society*. Syracuse, NY: New Readers Press.

Ligon, F., & Tannenbaum, E. (1990). *Picture stories*. White Plains, NY: Longman.

Magy, R. (1998). *Working it out: Interactive English for the workplace*. Boston: Heinle & Heinle.

Newman, C. (2000). *On-the-job English*. Syracuse, NY: New Readers Press.

Ringel, H. (2000). *Key vocabulary for a safe workplace*. Syracuse, NY: New Readers Press.

Robinson, C., & Rowekamp, J. (1985). *Speaking up at work*. New York: Oxford University Press.

Wrigley, H. S. (1987). *May I help you? Learning how to interact with the public*. Reading, MA: Addison-Wesley.

FIGURE 1. Worker-Centered ESL and Literacy Textbooks

mainly for introductory purposes but found that customized materials provided the best practice.

Teachers employed a rich variety of customized materials in the classroom. Company newsletters, work forms, paycheck stubs, training manuals, tools, parts, and safety equipment met the communicative needs of the workplace. To use these materials accurately and effectively, teachers studied the workplace structure and associated responsibilities of workers at the companies to which they were assigned. Teachers were dedicated to delivering customized classes that met the needs of the workers and were therefore willing to learn about all aspects of the workplace.

Attention to Additional Learning Needs

Although the WEP was initially set up to offer ESL and GED classes, union officials, business partners, and WEP staff quickly realized that workers also needed other types of classes. The need for basic math skills surfaced at one factory that was implementing statistical process control for quality assurance. Another company cited health and safety training as an immediate need. Other companies wanted team building and conflict resolution. To develop these courses, we, along with the teachers, conducted needs assessments that targeted the gaps in these areas. The resulting classes were open to all workers, not just nonnative English speakers.

When the WEP was first implemented, neither we nor the director had predicted that we would end up teaching such a variety of work-related courses. The program's final report (Boyter-Escalona, 1998) lists 18 course types offered at the various partner companies. The educational staff had experience with ESL and basic skills but not with conflict resolution or time management, even though these were the kinds of courses employees and employers wanted.

Consequently, with the help of human resource personnel and union staff, the WEP sought out materials covering these special topics, developed curricula, and trained teachers, most of whom had exclusively ESL experience. The willingness of program staff and teachers to incorporate these other courses into the language curriculum was critical to the WEP's existence and credibility. The courses attracted members of the workforce who otherwise would not have benefited from the program. Furthermore, the WEP became valuable to companies that did not regard their employees as being in need of basic skills. Also, in these special courses teachers could facilitate workers' reading and writing skills within the context of workplace topics.

Especially prevalent among the workers was the notion that they needed computer skills for professional mobility and personal growth. According to Soifer et al. (1990), computers benefit a workplace program because they

1. motivate learners

2. make writing, revising, and editing easier

3. engage learners in gathering and organizing data and using the information to solve problems in math science, and social studies

4. acquaint adults with the technology that has become so prominent in society and the workplace (p. 110)

The union solicited donations of computers and software, and provided a laboratory for the WEP's computer course. The curriculum, entitled *Workplace Communication and Computer-Assisted Learning* (Garcia & Sharma, 1995), listed learning objectives that included improving the reading and writing of documents, reading charts and graphs, creating spreadsheets, and expanding computer and work-related vocabulary. This curriculum guide integrated workplace topics with computer applications, such as filling out supply order forms, writing vacation requests, filing accident reports, and tallying hourly production using statistics from the factory. These classes were popular with both native- and nonnative-English-speaking workers.

◈ PRACTICAL IDEAS

The implementation of educational programs at factories offers a variety of challenges and caveats. WEP staff learned through experience, and a few mistakes, about the importance of building partnerships, addressing the needs of all partners, and being flexible enough to offer a multiplicity of courses. Over time, the WEP became so valued by the partner companies that, after the program ended, some of the companies continued to support educational classes for their employees. The union has also continued to offer several courses initiated by the WEP. This success did not come easily. WEP staff avoided some pitfalls by taking advice from other workplace education programs and the union. They learned other lessons through trial and error. Based on the experience, I have identified six criteria for running a successful workplace literacy or ESL program. These six principles contributed considerably to the WEP's overall success and longevity.

Offer Something for Everyone

This point cannot be overemphasized: All workers within the same factory or workplace must have access to some kind of educational program. In factories where workplace ESL is the main offering, the company should also offer courses for native English speakers, such as math, GED test preparation, team building, and workplace health and safety. Offering something for each worker ensures fair treatment for everyone and reduces opportunities for conflict.

Involve All Constituents

All members of the workplace community should be involved in planning the educational program. Company management at every level, human resources personnel, shop floor managers, and shop floor workers give valuable input to the educational program administrator. If a labor union represents any branch of the workforce, it, too, should be contacted and involved in program planning. Furthermore, educational program staff should not assume that management or the union will take responsibility for disseminating information about the educational program. Educational staff must make sure that advertising flyers are made and distributed and that employees are neither uninformed nor misinformed.

The WEP found that a partnership model in the form of advisory boards was the best way to ensure the involvement and representation of all constituents in the workplace. The advisory board assisted and advised the program at every level of its development, from needs assessment to evaluation. The board was also important as a reviewer of proposed curriculum, as the WEP learned when, at one company, the health and safety curriculum caught the attention of company managers. The managers thought that the curriculum blamed accidents on the company and did not place the responsibility of avoiding accidents on the workers. When the curriculum content was passed on to the chief executive officers for review, the WEP was suddenly threatened with complete removal from the company. Only after revising the health and safety curriculum to include company policy and promising to give copies of all curricula to company management was the WEP allowed to continue providing courses at that company.

Remain Politically Neutral

In most workplace situations, the educational program staff members are recent additions to a group of people who have known each other for a long time. Some workers reported having been employed at the same company for more than 20 years. At such companies, employees and employers have had a long relationship that might not always have been pleasant. Educational program staff, who are new to a company's group dynamics, need to be aware of the potential for animosity among the various constituents. In fact, educational staff may not want to implement a program in a company that is experiencing severe infighting or other problems until the conflicts are resolved. If they are not, educators may find themselves caught up in a longstanding political battle for which they are not prepared. Workplace teachers especially must be aware of such problems so that they do not become involved in company politics and can remain politically neutral.

Remain Visible

The WEP discovered that the best way to recruit new students and retain participating students was to maintain visibility on the shop floor. Teachers would often walk through the workplace on their way to class or to collect realia from production lines, and the other program coordinator and I also visited the shop floor regularly. The reason for this was the high turnover rate of part-time teachers. As much as the WEP tried to retain teachers in whom we had invested time and training, the unusual circumstances and part-time nature of workplace programs worked against this effort. Gaining workers' trust requires continuity of program personnel. Thus it was crucial for us to visit partner companies not only for advisory board meetings but also to develop curriculum, conduct plant tours for teachers, attend student recognition events, and conduct ongoing needs assessments.

Give Fair Compensation

As a way to reduce high teacher turnover rates, the WEP compensated teachers at their regular teaching rate for noninstructional activities: WEP meetings, advisory board meetings, plant tours, student recruitment activities, student assessment, and curriculum development projects. They were compensated at half their regular teaching rate for lesson planning. This arrangement was welcomed by many part-time teachers and augmented the success of the program.

Although educational program staff can ensure that their teachers are fairly compensated, they should also lobby company management for at least partial paid release time for participating worker students. The WEP's most successful educational programs took place at companies where workers were compensated for half of the time they spent in class. Fully and half-paid release time sends workers an important message about their value to the company and lets them know that the company is willing to invest in their basic skills.

Maintain Frequent Contact With Teachers

As discussed in the section Supporting Teachers, teachers in the workplace may feel isolated from their peers. WEP classes took place at odd hours, and partner companies were located over a wide geographical area, which added to the isolation of teachers. We maintained frequent contact with teachers over the telephone or in person by scheduling regular meetings and in-service training. We involved teachers in special projects such as curriculum development, task analyses, and the creation of the GWA. Teachers appreciated accruing additional hours by taking part in these projects, and their input was enormously valuable to the program.

◈ CONCLUSION

The WEP provided many benefits for the union, the partner companies, and, of course, the workers themselves. Participating workers increased their self-confidence by building their communication skills. Many workers reported feeling more comfortable using English on the job and outside the workplace (Boyter-Escalona, 1998). The union reported increased solidarity, a reduction in grievances, and improved relations with company management. Companies reported increases in production and reductions in communication barriers. The university that oversaw the grant also benefited because it could reach out to nontraditional learners and improve community relations. The university added validity to the WEP and fostered relationships with area businesses.

An interesting outcome of the WEP was that the provision of educational courses became a bargaining chip in contract negotiations. When the WEP was well established, the union organized two new companies; workers stated that one of the main reasons they voted for the union was that they wanted to attend workplace classes, particularly in ESL, GED preparation, and team building. These examples show the extent to which blue-collar workers desired improved communication and basic skills.

By the program's close, the WEP had succeeded fully in meeting its goals. Today, this ESP program continues to serve as a model of collaborative adult education and will likely continue to do so for many years to come.

◈ CONTRIBUTOR

Paula Garcia is a doctoral student in applied linguistics at Northern Arizona University. She taught and coordinated workplace ESL, literacy, and bilingual vocational education in the Chicago area of the United States from 1989 to 1996 and then taught EFL in Morocco. Her research interests include language assessment and listening comprehension.

CHAPTER 11

An ESP Program for Brewers

Liliana Orsi and Patricia Orsi

◈ INTRODUCTION

Performing successfully in the workplace can be stressful for professionals who need to use EFL to carry out workplace tasks and meet the challenges of continual career development in a competitive business world. ESP professionals can provide English language training that can significantly lessen the linguistic burdens of the workplace, thereby increasing the potential for success.

The ESP program described in this chapter was designed for professionals in the beer industry who needed to develop skills in English as well as acquire vocabulary for a specific purpose over a short period of time. The program's main objective was to prepare clients for an overseas seminar in which English would be used almost exclusively. Needs analyses showed that participants needed to improve and practice all four skills, especially listening and speaking in the situations that were likely to arise during the program.

◈ CONTEXT

We designed and delivered this in-house ESP program for one of Argentina's largest breweries. At the time, we were administering general English courses as freelance EFL trainers in the company's training department. We created materials especially for the program, incorporating an institutional video (Miller Brewing Co., 1990) from a U.S. brewery. This material followed the scope and sequence of the film and was later printed in a booklet (Orsi & Orsi, 1990).

Classes were delivered at the brewery's training center to small groups of adult native speakers of Spanish. The brewery supplied equipment, such as a television set and videocassette recorder, and photocopies (eventually bound copies) of materials. Most of the participants, aged 30–55 years, were engineers or technicians who had been selected to attend intensive professional training in the Netherlands. Some were already attending regular EFL classes but needed English language instruction specific to brewing to prepare them for this seminar.

A dynamic learning environment fostered interaction among students and teachers, as they studied materials together and shared their observations (Kaufman & Brooks, 1996). Teachers served as facilitators who assisted learners in acquiring the information and English skills they needed via team learning in the classroom.

Initially, we taught the course, but we later delegated teaching duties to three teachers who were new to ESP and to beer brewing. Selected based on their creativity and their ability to face the challenge of teaching in an unfamiliar content area, they at first reluctantly and then enthusiastically accepted the challenge of creating learning opportunities in an interactive, communicative classroom (Spada, 1990).

◈ DESCRIPTION

The program included 50 hours of instruction divided into seven units corresponding to the videotape on which the material developed for the program was based. Each unit was covered in three to six 60-minute sessions depending on students' abilities and needs. Classes, which had one to four participants, consisted of either two or three modules that incorporated intensive practice and follow-up in all skills. In line with company policy, classes met before work, at lunchtime, or after work, usually for 90–120 minutes. Participants enrolled in the ESP program were employees who had been selected to attend an intensive brewer's training program in the Netherlands.

Needs Analysis

We first carried out a needs analysis in order to find "identifiable elements" of the "students' target English situations" (Johns & Dudley-Evans, 1991, pp. 123–125). This analysis consisted of a language proficiency evaluation followed by an interview in which we identified specific, work-related needs for English and gained additional information about motivation and interest in the ESP program.

Placement Test

Even though most of the learners were already attending general EFL classes, we administered a placement test to determine their strengths and weaknesses in the English required to succeed in their seminar abroad. Because the learners would need to use all four language skills in the seminar, we evaluated each skill separately. Proficiency assessment took into account Krashen's (1982) principles of acquisition; thus, students were asked to work as fast as possible within a time frame that ranged from 5 to 10 minutes depending on the area evaluated.

We recorded the results graphically in a profile for each learner (see Figure 1). Along with the profile, we recorded comments such as these:

> Quite good proficiency, especially at the one-on-one conversation level.
>
> Uses language rather appropriately although not supported by solid grammar acquisition.
>
> He can recognize structures but can't produce them.
>
> Only basic grammar forms have been acquired and are properly used.
>
> He can perform better when reading familiar subject matter.
>
> Makes numerous mistakes of use and form. He possesses ability for informal communication but needs to upgrade his writing skills.

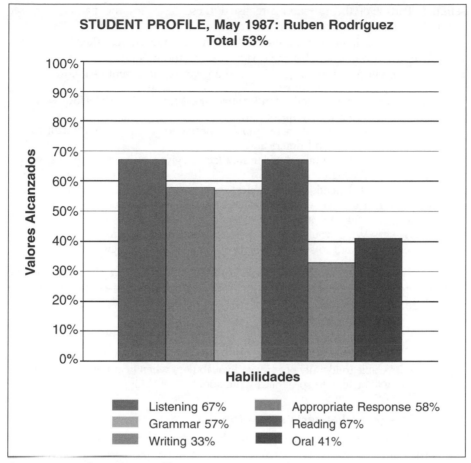

FIGURE 1. Sample Student Profile

Understands more than he can produce. He shows self-confidence and answers questions quite promptly although he needs to work on his fluency. He makes the typical non-native speaker mistakes ("the Jennifer's mother"). His vocabulary is limited and needs to be developed (*assist* for *attend; support* for *put up with*). He is reluctant to use past tenses except for the *be* verb.

Most of the learners were somewhat proficient in reading but needed to develop their listening, speaking, and writing skills. Their proficiency levels ranged from low intermediate to advanced.

Interview

For the interview, we designed a simple form for participants to complete in Spanish, their native language. In Part 1, they evaluated their English proficiency by identifying tasks they could and could not do in English. Although we already had information from the training department about the tasks required in the overseas seminar, Part 2 of the questionnaire listed these tasks for the participants (see Figure 2), who checked off their most urgent needs.

After the participant had completed the form, the interviewer asked questions in English about how comfortable each participant in the program felt when performing tasks in English. A typical question-and-answer session went as follows:

Interviewer: Can you understand native speakers of English? How well? What if they speak fast?

Learner: I get lost when they make contractions . . . . it is difficult to understand Americans.

Interviewer: Can you take notes from a lecture given by a native speaker of English?

Learner: It depends on the subject matter.

Interviewer: Have you ever been in an English-speaking country?

Learner: Only on vacation.

Interviewer: What would you like to gain from this program?

Learner: I want to survive the course and learn about new technology.

Once we had obtained a complete profile of the participants, we shared it with the participants themselves and with the instructors. Reviewing the profiles helped set realistic goals, such as the following:

At the end of the course you will be expected to follow a lecture, take notes, and ask questions.

You will be able to identify beer related vocabulary when heard or when read, and you will be able to speak and write about the stability of foam.

Course Design and Materials Development

Initial evaluations, needs assessments, and the resulting learner profiles showed that the participants' greatest need was to develop listening and speaking skills and to learn English vocabulary related to the brewing industry. The course was designed around an institutional video (Miller Brewing Co., 1990) from a leading U.S. brewery

---

I need to be able to:

☐ Answer faxes

☐ Answer e-mails

☐ Write reports

☐ Socialize

☐ Use the telephone

☐ Talk to native speakers

☐ Make a presentation

☐ Listen to presentations

☐ Understand aural media

☐ Other

FIGURE 2. Excerpt From Needs Assessment Form

that did not have any business interests in Argentina. The video, which was easily purchased from the company's training center, had been developed to introduce new employees to the characteristics of the company and its products. Because it had been created for native speakers of English, it included authentic material quite similar to what our program participants would be exposed to in their seminar abroad: ingredients, equipment, and procedures used in beer making; fermentation times; and quality control procedures.

Before designing the course materials, we contacted engineers in the production department to make sure we had selected the correct content. Our two volunteer consultants were the production manager/brew master and his assistant, an engineer in the area of maintenance who was familiar with the equipment and technology of brewing. The second volunteer later became the maintenance manager of the plant. These two employees played an important role in assuring us that we had selected and treated content appropriately and guiding us in highlighting the company's most pressing needs. With them, we negotiated goals and planned the program based on the premise that successful learning processes are constructive and autonomous and take place in authentic situations. We also consulted previous participants in European training seminars to obtain feedback on their experiences.

Besides the video, we needed reading texts that would provide practice aimed at triggering discussion on topics that might arise during the seminar workshops or as part of social-business exchanges with colleagues from other breweries. The reading passages, which came from brewers' magazines (e.g., *American Brewer*), included want ads and articles such as "Tackling a Big Job" (1915) and "Normal Beer and Nonalcoholic Substitutes" (1915). Our challenge was to design appropriate activities and materials that would help our students understand complex reading pieces that were often beyond their language proficiency. Understanding subjects and objects consisting of heavy noun phrases, syntactic markers of cohesion, and the role of nontechnical vocabulary in technical texts is problematic even for students who have mastered the technical terms (Cohen, Glassman, Rosenbaum-Cohen, Ferradas, & Fine, 1979).

To address important grammar points during the course, we used a grammar book. We also incorporated realia such as labels from beer bottles and ad clippings.

The course materials used basically the same content and followed the sequence in the video for all students regardless of English proficiency level. Some classes were multilevel in terms of English proficiency, with participants sharing a similar degree of background knowledge. Occasionally the participants lacked proficiency in English but possessed exhaustive knowledge of the subject matter. This background knowledge seemed to even up their performance, providing self-assurance and raising self-esteem. Our thinking along these lines is supported by Rosenblatt (1994) and by Floyd and Carrell (1994):

> The residue of the individual's past transactions in particular natural and social contexts constitutes what can be termed a linguistic-experiential reservoir . . . . Embodying funded assumptions, attitudes, and expectations about language and about the world, this inner capital is all that each of us has to draw on in speaking, listening, writing, or reading. We "make sense" of a new situation or transaction and make new meanings by applying, reorganizing, revising, or extending public and private elements selected from our personal linguistic experiential reservoirs. (Rosenblatt, 1994, p. 1061)

> Prior background knowledge of the content area of the text (content schemata) significantly affects comprehension of that text. . . . Background knowledge relevant to reading comprehension can effectively be taught in the ESL classroom. (Floyd & Carrell, 1994, pp. 309, 322)

Tapping our learners' subject matter knowledge and training, we believed, would ease the way to improving performance and provide situations in which we could supply valuable *comprehensible input* (Krashen, 1982) that would later become part of the participants' productive English vocabulary and language skills (Vygotsky, 1980).

Based on these principles and using the material described, we developed seven units of English study, printed in a 89-page spiral-bound book (Orsi & Orsi, 1990) to accompany the video:

1. What's Beer?
2. Brewing, Step 1
3. Fermentation, Step 2
4. Aging and Filtering, Step 3
5. Genuine Draft
6. Pasteurization, Bottling, and Packaging
7. Review Questions, Vocabulary, and Crossword Puzzle

Each unit exposed learners to specific target language as well as practice and development of the four skills. We made sure students were aware of their own learning strategies (Gardner, 1983), needs, strengths, and weaknesses. Each unit included a wide variety of specially designed activities to go along with the video. To assist teachers who were new to the brewing industry and did not feel confident about teaching the English of an unfamiliar content area, we included an answer key at the end of the textbook.

Sample Activities

Each class incorporated individualized instruction and practice in each of the four skill areas. Sample activities for each area are described below.

Listening

Participants needed enough exposure to English to feel comfortable with the language (cf. the *affective filter hypothesis;* Krashen, 1983) when listening to lectures on such topics as the stability of foam, types of beer, and beer ingredients. They also needed practice in getting the main idea, taking notes, and summarizing. Target vocabulary was provided by the video, which described the different stages of the brewing process along with the history and evolution of beer. Listening activities to accompany the video included the following:

- Listen and take notes on the characteristics of the different types of beer.
- What are the differences between ales and lagers in body, taste, bitterness, and color?
- Which are the different types of ales and lagers?
- What kind of beer do you make at this brewery?

- What differences are there in the ingredients used and the process followed?

- Are adjunct cereal grains added in all processes? Why or why not?

- How can you compare the Argentine beer market to the European market?

These questions triggered lively discussion, leading us to integrate speaking practice here. For example, adjunct cereal grains (i.e., rice or corn) are not added to all beers; in fact, they are prohibited by the *Reinheitsgebot,* the German beer purity law, which regulates beer ingredients. Discussion of topics like this interested our participants very much.

Other types of listening activities included multiple-choice and listen-and-check items, for example,

Listen to the video and choose:

1. The ingredients for beer are
   a. water, hops, malt and corn.
   b. water, hops, malt and yeast.
   c. barley, hops, malt and yeast.

Listen and tick the statements that you hear:

1. True draft beer comes in cans and bottles as of 1986.
2. Most German beers are lagers.
3. Adjunct cereal grains are not used to produce German beers.
4. The Pilsners are second in popularity in the USA.
5. The Pilsners are somewhat lighter in color and body than the standard lagers.
6. *Light* does not mean *low-calorie.*

Reading

While attending the training seminar, participants would have to do a great deal of reading on new scientific information related to the industry. We therefore designed reading activities aimed at providing as much authentic practice as possible. Activities aimed at improving general comprehension and language development as well as problem solving also developed critical thinking:

> When there's indication that the text itself has lost its individuality and its information content has become integrated into some larger structure . . . knowledge makes understanding processes smart . . . general knowledge about (words, syntax, the world) anything makes the construction of representations possible . . . a text base is constructed from the linguistic input as well as from the comprehender's knowledge base. (Kintsch, 1988, pp. 951–953)

Reading activities included selecting right and wrong statements; matching words to their meanings; finding the closest meaning; answering information, open-ended, and opinion questions; and identifying the adverb corresponding to an adjective (e.g., *fresh-freshly*) or the verb corresponding to a noun (e.g., *fermentation-ferment*).

Speaking

In our initial needs analysis, we discovered that during the training seminar our learners would be expected not only to speak about beer in professional terms but also to speak informally and with humor about beer and various other matters on occasion. Consequently, the course included instruction in telling beer-related jokes in English.

Because the sponsor of the training seminar was a very well known European brewery, we came up with two versions of a joke capitalizing on the rivalry between U.S. and European beers. The general social version was, "Do you know the similarity between American beer and a canoe?" "They are both damn close to water." For the men-only version of the joke, *damn* was replaced with an obscenity in its *-ing* form. We also introduced other obscenities and explained that although language of this variety is not socially problematic in Argentina, in other cultures the use of obscenities in conversation with certain people is considered offensive. The learners rehearsed the joke until they felt comfortable with it. Some of the participants loved the idea of being able to tell a joke in English because they were good at telling jokes in their native language. Others declined to tell jokes. We exploited this opportunity to discuss cultural differences; as Gatbonton and Tucker (1971) point out, "misunderstandings due to the application of inappropriate values, attitudes, and judgement (would not take place) . . . if cultural barriers could be overcome" (cited in Floyd & Carrell, 1994, p. 312).

The participants also learned to ask questions about subject matter that they did not understand and to discuss trends in marketing and packaging. These speaking activities were ongoing, and classes were very interactive. In addition, each unit included a section called "Something to Talk About," which created the opportunity for teachers to help students verbalize what they already knew about the topic and establish relationships between prior knowledge and new material (Hayes & Tierney, 1982, cited in Floyd & Carrell, 1994, p. 312).

Kintsch (1988) has claimed that "a text base is constructed from the linguistic input as well as from the comprehender's knowledge base" (p. 953). He also argues that, when listening and reading, individuals comprehend discourse not only by analyzing the units in the discourse (e.g., syntax, word order, aspect, form) but also—and most importantly—by imagining a situation described by the text but re-created by the individual depending on the influence of general world background knowledge, beliefs, prejudices, perceptions, and lexis. This re-creation is helped more by long-term memory, inference, and retrieval capabilities than by short-term conscious memory (Kintsch, 1994). Tied to these claims, Smith (1985) has stated that "students' ability to predict and comprehend depends on their individual theory of the world; their ability to ask the relevant questions necessary to find the answers will lead to comprehension" (p. 80). Thus, to help tie the participants' English learning to their world, we used excerpts from *American Brewer*, some dated as far back as 1915. Want and offer ads (see Figure 3) served as reading and discussion material regarding changes in education and experience required to perform jobs in the beer business. Participants rewrote the ads, updating them for publication in a newspaper today, and we discussed their choices for revision.

Readings on ethical issues concerning drinking, regulations for underage drinkers, and historical and social settings raised the opportunity for the participants

> **Wanted.** A lively, energetic Manager for a first-class modern Brewery located in mining region of Western Pennsylvania, being the only brewery in the County, containing 50.000 to 20.000 barrels and business showing a large increase monthly, with reputation of making finest beer in the State. Party must be in a position to take stock in the company. None but those meaning business and having business need apply. Address: N.B. 11, c/o American Brewer 200 Worth St., New York City.

> **Brewer Wishes to Make a Change.** Brewer with fourteen years experience, wishes to make a change from present position, where he has been employed as brewmaster for the past eight years, and is not obliged to leave. Holds first grade diploma from United States Brewers Academy. Brews lager and ale, and has made sparkling ale a special study. Advertises from one of the leading families of brewers in the country.

FIGURE 3. Sample Ads (*American Brewer,* 1915)

to discuss cultural matters and differences, which was challenging for those who had never traveled abroad. We chose texts and supplied prompts such as those shown in Figure 4.

Writing

During the training seminar abroad, learners would be expected to write short notes to colleagues from other countries as well as a final report on their own company. The course therefore included instruction in writing reports and informal thank-you and invitation notes.

A read-discuss-write section in each unit integrated reading, discussion, and writing practice. Prompts included the following:

- Discuss beer versus nonalcoholic substitutes.

- Write an advertisement for a beverage.

- After reading ads in the 1915 issues of American Brewer, compare and discuss the beer industry today and the beer industry then.

- Write a paragraph for or against Prohibition based on your reading of the article "Tackling a Big Job" (1915).

Vocabulary Building

At the end of the textbook was a brewing vocabulary crossword puzzle (see Figure 5) that students could complete as they learned new vocabulary. In addition, the book included a glossary of terms in English, and students were to complete it with equivalents in Spanish.

Think while you read: What do the terms *dry* and *shy* mean in this context? Why is the word *evangelists* between quotation marks? Should there be a drinking age? Do you think Prohibition leads to excess?

Tackling a Big Job

The prohibitionists in Chicago are planning an anti-saloon campaign in that city, with a view to forcing a vote on the license question at the municipal election next Spring. . . . This means that 60.000 signatures must be obtained. . . . There was a similar movement in 1908, but the petition was "shy" 5,000 names when the time limit for filing it arrived. There is no general alarm as to the prospect of Chicago becoming "dry," although the practical politicians dread the complications that will ensue in connection with an anti-saloon campaign. The success of such a campaign would put out of business 7,000 Chicago saloons, licensed under the Illinois Local Option Act of 1907. Suppose that the so-called "evangelists" and others who oppose the saloon would succeed in convincing the voters for a period sufficient to bring about no license, what would be the result? Would temperance be the gainer? . . . . Know that "prohibition" leads to excess, not temperance, and that the hysteria of "temperance" crusaders can be called nothing better than honest error. When will the would-be reformers of humanity learn that if a man wants to drink he will drink?

Source: "Tackling a Big Job" (1915).

While you read these excerpts, do the following.

Read-Think-Discuss:

How was life different in 1915?

What was going on in the United States then?

Have values changed? For the better?

The Massachusetts Woman Suffrage Association disclaims that it is in favor of Prohibition, and declares that its position is one of neutrality, ... At the same time the Anti-Suffragists deny that they are allied with "the liquor interests" . . . .

ARIZONA. Wine for sacramental purposes will not be shipped into Arizona by railroad and express companies until the new Prohibitions octopus has been tested in the courts of this state . . . .

As the peddling of liquors, or so-called Bootlegging . . . . Peddlers of liquor or bootleggers are not to be regarded as coming within the class of unintentional violators and should be arrested and reported for prosecution whenever found selling liquor in such manner . . . .

The Phoenix police have their hands full in prosecuting soda water sellers who put a "stick" into their soft stuff; grocers also give lots of trouble as they begin selling whiskey in cans bearing tomato labels . . . .

Source: *American Brewer* (1915).

FIGURE 4. Sample Speaking Prompts and Texts

■	1 B	E	E	R	■	2 R	I	C	E	■	3 L	A	U	T	E	R
4 L	A	G	E	R	■	5 M	A	S	H	■	6 B	O	T	T	O	M
7 C	R	O	W	N	C	O	V	E	R	■	K	E	R	N	E	L
9 P	I	L	S	E	N	■	10 H	O	P	S	■	11 A	L	L	O	W
12 B	O	C	K	■	13 T	U	N	■	14 A	G	I	N	G	■	■	■
■	■	■	15 C	L	A	R	I	F	I	E	D	■	16 B	R	E	W
■	■	17 C	O	N	V	E	Y	E	D	■	18 C	O	O	L	E	R
19 S	T	O	U	T	■	20 A	L	E	■	21 C	A	N	■	22 K	E	G
23 B	R	E	W	I	N	G	■	■	■	B	R	E	W	E	R	Y
25 H	O	P	S	■	26 A	L	E	■	27 U	P	■	28 D	R	A	F	T
29 V	E	S	S	E	L	■	30 M	I	L	L	31 F	U	L	L	■	■
32 S	I	L	O	■	33 W	H	I	R	L	P	O	O	L	■	■	■
■	■	■	34 A	R	O	M	A	■	35 T	W	O	■	36 W	O	R	T
37 Y	E	A	S	T	■	38 W	I	N	E	■	39 B	R	E	W	E	R
40 B	O	T	T	L	E	■	41 I	N	G	R	E	D	I	E	N	T
42 P	U	R	I	F	Y	■	43 C	O	O	L	■	44 C	R	O	W	N
45 P	U	M	P	E	D	■	46 T	R	U	B	■	47 H	E	A	T	■

Sample clues:

1. The beverage of moderation. (beer)
10. Gives beer it bitterness. (hops)
17. Transported. (conveyed)
20. Type of beer made with top fermentation yeast. (ale)
22. Barrel. (keg)
29. Tank. (vessel)
34. Pleasant smell. (aroma)
36. Sugary liquid. (wort)

FIGURE 5. Crossword Puzzle Key and Sample Clues

DISTINGUISHING FEATURES

Autonomy and Independent Learning

Nunan (1995) expands on and supports the need for a learner-centered curriculum "to help learners develop strategies that will help them learn" and be actively involved in "setting their goals and interests, making choices about tasks, content, and the direction of the learning" (pp. xi–xii). In trying to make our program successful, we created a caring and sharing atmosphere and tried to maintain it while aiming at increasing learners' autonomy. We sought to foster an ability to take charge of one's own language learning (Holec, 1981) especially because, once in Europe and on their own, the participants would have to function in English without the help of their instructors.

One incident that shows our success in encouraging independent learning was when students spontaneously brought to class sample charts for their mashing formula and brochures that described the process at their brewery. These contributions led to active and lively discussions comparing the different processes, the variety and quantity of ingredients, the appropriate machinery to use in different situations, and the steps of the brewing process (e.g., using rice instead of corn as an adjunct cereal grain, or seeing filtering as part of the fermentation stage or as a separate stage of the brewing process).

Focus on Specific Needs

This program was typical of ESP because it was designed to address the specific needs of a particular group of learners (Hutchinson & Waters, 1987). The material was challenging for the beer professionals participating in the program because, even though most of them knew a great deal about the subject matter, they had had no exposure to it in a foreign language. In other words, they knew how to make beer in Spanish, but they had never done so in English. Although our program focused mostly on preparing participants for the training seminar in the Netherlands, it also paved the way to making them bilingual in "beer talk." In the future, they would have to continue and strengthen their professional relationships with Dutch colleagues who would visit Argentina, and they would need to continue to communicate with them throughout the year by telephone and fax.

Teacher-Student Teamwork

Teaching in the ESP program was a formidable task for the EFL teachers, who accepted the challenge of teaching within a content area totally unfamiliar to them. Teachers were briefly trained but mostly learned from experience. Some of the teachers had to be encouraged to accept the challenge because it was their first ESP experience, and they were a bit reluctant and insecure at first. Later, however, all of them felt the experience was highly rewarding.

What made the experience easier for the English teachers was that the program was a team effort: Teachers offered the language instruction, and students helped supply the content. This joint responsibility greatly enhanced the learning experience (Drucker, 1992). The program fostered collaboration and involvement by drawing on each participant's knowledge and personal experiences. The answer key at the end of the textbook also proved especially helpful for teachers who were new to the brewing industry.

A Program Model for Other Industries

This ESP program for brewers serves as a model for ESP programs in other industries, especially for colleagues new to ESP in workplace contexts. The video was particularly useful for exposing students to authentic language and, in this case, guided the administration of content.

◈ PRACTICAL IDEAS

Designing an ESP program within an unfamiliar professional field poses a challenge to the ESP practitioner. The work is hard and time-consuming, although undoubtedly rewarding for teachers ready to accept the challenge. We offer three pieces of advice to colleagues undertaking similar projects.

Use Authentic, Appealing Materials

Materials should be authentic and should appeal to the learners. Samples of genuine English that learners will need and that cover topics of interest to them will yield better results than English texts created for nonnative English speakers on topics that interest only English teachers.

Form a Partnership

Program instructors need to cooperate with all stakeholders, as a single stakeholder seldom knows enough to do all of the work alone. Creating a learning experience partnership between instructors and participants proved crucial for the construction, implementation, and completion of our program. Teachers and students were involved throughout, from needs assessment, to materials development, to final evaluation.

Use Existing Materials If Available; Create Materials When Necessary

We were fortunate to locate an institutional video from a similar brewing company to use in our course. The fact that instruction was not based on textbook English motivated teachers and students and generated greater participation from both groups.

We also wrote new materials for language instruction and placed them in a hardcover booklet that was printed inexpensively by the brewery. Having such a booklet made management of the learning materials convenient for both teachers and language learners.

◈ CONCLUSION

The success of our ESP program was due to its collaborative, learner-centered nature from inception to delivery. First, we tapped our students' professional knowledge of brewing to create materials in English that were tailored to the students' learning needs. Students and teachers built a relationship based on mutual respect: Participants did not expect teachers to know about beer making, and teachers did not expect participants to be fluent in English. Each group contributed what it knew, thus making the learning a joint effort.

In addition, the brewery managers showed that they valued our professional ESP approach: They assisted us in course design, made positive comments during the program, and published our textbook twice and distributed it for use at other plants in Argentina. Our publication also served as a sample to convince future clients of the need to design materials specific to their industries.

Students performed tasks that resembled those done in real-life situations, allowing them to reflect on their professional performance, and materials were developed especially to support learning. Working with content that was familiar and relevant to the industry and their jobs undoubtedly raised our learners' self-esteem and kept their level of interest and participation high.

In summary, the program was highly successful in preparing course participants for the challenges they were to encounter abroad. During our program, learners made the most of their previous knowledge and acquired additional knowledge along the way. Instructors balanced content and language, and employed effective classroom strategies that made appropriate use of authentic materials. In informal interviews after their return from the Netherlands, the learners reported that they felt comfortable using the English they had learned and regarded the experience as a fruitful one. An unexpected result was that the overseas experience led to improvements in interactions between workers who had received training abroad and their coworkers in Argentina who came from other countries.

The material designed for this ESP program has become an important part of the brewery's library. It is now being used as an introduction to the brewing process for newcomers to the beer industry at other facilities owned by the client brewery.

◈ CONTRIBUTORS

Liliana Orsi holds a teacher's degree from the Presbítero Sáenz Institute, in Lomas de Zamora, Buenos Aires, Argentina, and is a candidate for the *licenciatura* at the Universidad Nacional del Litoral, in Santa Fe, Argentina. She is academic director of Rainbow, a private language school and ESP consulting agency employing 25 practitioners in Lomas de Zamora, teaching on-site and administering ESP programs at various workplaces in Argentina since 1984.

Patricia Orsi holds a graduate degree in demography and tourism from John Kennedy University, in Buenos Aires, Argentina. She is executive director of Rainbow.

Liliana Orsi and Patricia Orsi were the TESOL ESP Interest Section members-at-large for 1997–1999 and have been frequent ESP presenters at conferences. They founded Argentina TESOL's ESP Interest Section and encourage networking among professionals in the field.

CHAPTER 12

An ESP Program for a Home-Cleaning Service

Patricia Noden

◈ INTRODUCTION

This chapter describes an ESP program for a home-cleaning company located in Northern Virginia, just outside Washington, DC, in the United States. The primary goals of the program were to raise the level of the cleaning staff's English skills in order to improve customer service and avoid language-related problems that lowered productivity. The chapter describes needs assessment, program design, logistics, curriculum development, instructional methodologies, feedback, and reinforcement mechanisms. It also describes challenges the program faced, some of which were particular to this program but most of which are common to ESP programs in the service industry.

◈ CONTEXT

The Service Industry

In the Washington, DC, area, most businesses employ nonnative English speakers for front-line service jobs at hotels, restaurants, and cleaning services. The area typically has low unemployment rates (e.g., 3.2% in 1998 compared with 4.5% nationally; U.S. Bureau of the Census, 1998). Home-cleaning services are part of a larger industry classified as building cleaning and maintenance services. According to the Virginia Employment Commission (n.d.), maid and housecleaning services were the two fastest growing jobs in the Virginia part of the Washington, DC, metropolitan area between 1996 and 2000, with growth estimated at 16%. In this environment, employers usually must hire workers with limited English skills—in fact, far more limited than managers prefer.

The Client

The ESP program described here involved a company providing residential cleaning services in the Virginia suburbs of Washington, DC. When the ESP program began, the client's 50-person cleaning staff consisted almost entirely of Spanish-speaking women from Mexico, Central America, and South America. The office staff included a (bilingual) manager in charge of day-to-day operations and employee supervision. In the beginning of the program, two bilingual employees were in charge of distributing the day's assignments and handling scheduling conflicts and other problems on the day's agenda. Later, a native-English-speaking employee performed this function. Native-English-speaking staff took care of sales, the scheduling of visits with homeowners, customer service, billing, accounting, and other office functions.

The company's cleaning staff needed to speak, understand, read, and write English. Communicating with customers and office staff required speaking and listening comprehension skills. Employees needed reading skills to understand, for example, information about the houses they were scheduled to clean—including the location of and directions to the home, the type of service to provide, special requests from the customer, and internal administrative information. Cleaning staff also had to be able to read notes left by customers at the house, which usually listed requests, gave instructions or information, or asked questions; and to write notes to customers informing them about the services provided, scheduling, and other internal administrative matters.

Having native-English-speaking office staff and customers along with a cleaning staff with limited English proficiency, the company often experienced operational and customer service problems. These problems cropped up unexpectedly, demanding immediate attention and disrupting managers' day-to-day activities. They also lowered revenue and productivity, created unnecessary expenses, negatively affected customer service, and in extreme cases prompted customers to change cleaning services.

In one incident, for example, a customer left a note instructing the cleaning team not to pull the drapes in the living room because the rod was broken. A team member pulled the drapes, and the rod tore out of the wall, damaging property on its way down. The customer was upset, and to right the matter, the company refunded that day's fee, paid for the damage, and convinced the customer not to take her business elsewhere. In another incident, a cleaning team that had become lost on the way to a house called the office for directions and was unable to communicate with the native-English-speaking office staff. The bilingual staff member with knowledge of the area was unavailable, and the team was delayed for more than an hour. Because the team's schedule was full that day (as it was most days), the delay not only required rescheduling of the cleaning of that home but also disrupted the rest of the day's schedule. Also reported were pets let loose, use of incorrect cleaning products, the cleaning of rooms that had not been requested, and other problems. Incidents like these—and problems with other simple communicative tasks—affected productivity. An office staff member commented,

> I really get uncomfortable when the customer calls the office with a problem occurring right then. The cleaning crew can't explain it in English, so I have to go get [the bilingual manager]. We need him because he can talk in Spanish to the crew and also in English to the customer. If I have to get someone else who knows Spanish but not so much English, I'll have to put them on the phone to talk with the crew, then get them to tell me what the crew said, and then get back on the phone to talk to the customer. It gets a little crazy sometimes.

The ESP Provider

The client contracted my company, Step-Wise Training & Consulting, a multiservice firm with headquarters in Falls Church, Virginia, to design and conduct a language training program to reduce problems caused by the cleaning teams' lack of English proficiency. Included in the contract were provisions for designing, conducting, monitoring, and evaluating the program. Later, special courses for select groups, such as team leaders, were added as needed. The program began with a 12-week pilot and was extended for more than $3\frac{1}{2}$ years.

◈ DESCRIPTION

The program was developed in phases, which are best illustrated through a process of training design known as the *critical events model*. According to Nadler (1988), the developer of this model, training design involves a series of critical events with accompanying activities:

- identify needs
- specify job performance
- identify learner needs
- determine objectives
- build curriculum
- select instructional strategies
- obtain instructional resources
- conduct training
- (throughout the program) evaluate and give feedback

Performing a Needs Analysis

The needs analysis involved examining the company's business processes, the tasks necessary to achieve organizational objectives, and the communicative functions related to the language problems. The result was a detailed description of expected versus actual results. As Robinson (1991) points out, although needs assessment is often referred to as a singular entity, it actually consists of many activities. I distributed written questionnaires to staff members, interviewed key personnel, reviewed documents and videotapes, and held numerous meetings and discussions with the president, management and office staff, and employees.

Determining Training Goals and Objectives

The overall goal of the program was to reduce or eliminate the impact of language-related work problems. Training objectives focused on job-related language functions and desired business outcomes, mirroring the underlying philosophy behind ESP stated succinctly by Kavanaugh (1999): "The desired outcome isn't fluency but productivity" (p. 2).

Positions on the cleaning teams required different language functions, and the objectives reflected this. Requirements for team leaders, for example, were more demanding than those for team members. Objectives thus reflected different starting proficiency levels and company requirements (see Figure 1). Some team members were at the single-word stage of English language development. For them, objectives included learning the names of company supplies and formulaic constructions, such as introducing oneself as a representative of the company, greeting a customer, responding to greetings, using expressions to begin and end conversations, asking for repetition or clarification, explaining a lack of understanding, and explaining the need to ask for assistance.

The training plan grouped objectives discretely by level, and, more useful from the company's perspective, by function across levels (see Table 1). Objectives at higher levels of proficiency were less prescriptive regarding specific language, as higher level language users can generate their own language and no longer need to rely on memorized, formulaic expressions. This difference also reflected current knowledge about learning strategies in second language acquisition, namely, that "the student has to become uninhibited in extrapolating from one known use to other possible uses in analogous situations" (Rivers & Melvin, 1977, p. 167).

At advanced levels, too, company policy and philosophy were integrated into the objectives. For example, the company required cleaning teams to distribute comment cards to collect customer feedback. Team leaders learned the language necessary to motivate customers to complete the cards. Team leaders also learned the language to use for sales opportunities and for establishing and maintaining customer goodwill. Other functions addressed in the training were

- understanding information on the field sheets, asking for clarification, explaining and discussing items and work procedures

- understanding spoken and written directions (especially with regard to finding particular locations)

- understanding and being able to discuss common customer service issues

TABLE 1. OBJECTIVES FOR THE FUNCTION *DISCUSSING SCHEDULES*, BY LEVEL

Proficiency Level	Objective	Sample Language
Beginning	Identify current location	"We are at [address]." "We are in the car going to [address]."
	Indicate progress	"We are almost finished." "We will be 20 more minutes."
Intermediate	Notify staff of problems	"We can't unlock the door." "We are lost."
High intermediate	Notify staff of more complicated problems	"The sheet said the key would fit in the back door, in the deadbolt. It doesn't."
Advanced (team leaders)	Be able to handle a variety of situations involving more complicated vocabulary and conditions	"I tried jiggling the knob, but it still won't open." "She says she overpaid last time and has a $50 credit coming to her." "She needed us to finish by 11:30 a.m., or we would have had to reschedule."

Sample Speaking Objectives for the Function *Discuss Service Issues*

Function	Sample Language

Team *members* at the intermediate level of English proficiency will be able to do the following:

• Explain service provided	"We're going to change the sheets now." "We'll be going into the kitchen next."
• Ask questions	"Do you want the trash put out in the garage?"
• Understand instructions	"Please leave this door shut." "Please don't let the cat out." "Please don't wash these sheets on hot."
• Understand information	"I left the check on the counter." "I will fill out the comment card and leave it for you next week."
• Inform the customer about lack of understanding and provide solution	"I'm afraid I don't understand. Let me go get my team leader to help."

Team *members* at the high-intermediate level will be able to do all of the above plus the following:

• Explain procedures and ask for customer agreement	"We plan to start upstairs and work our way downstairs. Is this all right with you?" "We use degreaser on this; we can use [another product] if you prefer."

Team *leaders* at the high-intermediate or advanced level will be able to do all of the above plus the following:

• Discuss company policy, such as cleaning procedures, scheduling, and payment	[language from company documents]
• Handle complicated situations such as rescheduling and payment issues	[actual situations from company history]
• Respond to instructions having complications	[actual situations from company history]
• Explain the need to call the office; discuss alternatives for calling (e.g., employee or customer calls, now or later)	"I'll need to call the office to check on this." "The customer service staff at the office will be happy to discuss that with you."

Sample Objectives for Reading Customer Notes

- Read notes left by customers
- Request help from office staff for unknown portions of notes

Sample Objective for Writing Customer Notes

- Write notes that describe, explain, inform, and (at advanced levels) persuade customers concerning the service provided, payment and scheduling issues, problems and opportunities

FIGURE 1. Sample Objectives

- explaining feedback cards to customers
- responding to feedback (i.e., practicing good customer service)
- being able to explain to customers in conversation or in writing common issues such as breakage and damage, and variations from customers' requests
- discussing work procedures with office staff one-on-one and during staff meetings
- ordering and refilling supplies
- explaining the reasons for and the state of broken vacuum cleaners and other company materials or equipment
- asking for assistance and responding to requests for assistance
- negotiating misunderstandings in person and on the telephone resulting from language or technological problems (e.g.,poor reception on cellular telephones)
- discussing paycheck issues (e.g., late payments, incorrect amount, lost checks)

Identifying Learners' Needs

As in many workplace ESL programs, expense was a major issue. The company did not have the resources for individual, one-on-one oral testing; therefore, I administered a simple written test that identified approximate levels of proficiency, established a baseline for evaluation purposes, and guided materials development and choices of instructional methods.

Of the 37 employees who were pretested, 25 scored at the beginning or very low intermediate level; 6, at the intermediate or high intermediate level; 4, at the advanced level; and 2, at a literacy level. Noting the employees with literacy-level scores, I discussed with the company owner whether the training would benefit these workers or whether they should be referred to an external language program. Because these employees were the most in need of English instruction and because the class would, at least at first, involve a narrowly restricted amount of English, we decided to include them. Even within the categories of literacy, beginning, intermediate, and advanced, skills and abilities varied. Some of the intermediate-level and lower advanced-level employees could handle some spoken English, but their knowledge of English spelling, grammar, and syntax approached that of the beginning-level employees. Some of the beginning-level learners had almost no English vocabulary; others had quite a bit but still could not generate enough meaningful English to answer simple questions. Some of the intermediate-level learners could write quite well, whereas a few of the more advanced-level learners struggled with spelling and with basic rules for subject-verb agreement.

Three learners who said they could not write during pretesting actually could; they hid their skill because they were ashamed of their inability to deal with the English spelling system and grammar rules. Generally, however, the responses on the pretest gave a good indication of the learners' level, the content areas I needed to cover, and the learners' interests. This information, coupled with a prioritized list of topics and functions gathered during the needs analysis, guided curriculum development and choice of instructional methods.

Program Delivery

Class was held once a week from 7:30 to 9:00 a.m. On that day, cleanings were delayed until after class. The class met in a large L-shaped room where staff meetings and work-related training sessions were held, supplies were stocked, and company information was posted. In one part of the room were shelving units on which cleaning crews kept their supplies. In the other part of the room were several large tables and stackable chairs, which we used as classroom furniture.

Curriculum

Lesson planning was particularly challenging. A typical lesson plan had to account for different skill levels (literacy- to advanced-level); different lengths of enrollment in the class (new vs. returning vs. veterans); different attendance patterns (ranging from perfect attendance to sporadic attendance); and, at times, different job responsibilities (team leaders vs. team members). Because of the changing faces each week, offering consistently sequential lessons became impossible. Each lesson had to be self-contained.

With the learners, I developed a textbook based on freestanding weekly lessons, each focusing on a given function and containing review activities for new participants. The first lessons addressed the functional topics that the needs assessment identified as having the highest priority. For example, an analysis of company field sheets (containing the information cleaning crews received on each home they were scheduled to clean; see Figure 2) helped me understand the types of services provided, the situations requiring communication skills, and the language the company used. It also provided the content for functional grammar instruction.

Each lesson incorporated speaking, listening, writing, and reading as well as a few minutes devoted to pronunciation. The participants also learned new vocabulary, including words, common expressions, business terminology, and idioms. Pragmatic language use was covered as it related to the content. Grammar was integrated into the lesson via a whole language approach (i.e., exposure in a dialogue supplemented when necessary with overt instruction on essential rules, e.g., conjugating verbs and learning irregular past forms). All learning activities took place within the context of meeting the training objectives for a particular unit. I adjusted the content, teaching methodology, and activities as necessary and continually recycled information to reinforce learning and to incorporate new members of the class.

Instructional Strategies and Resources

Instructional techniques almost exclusively involved those recommended in the literature on multilevel classrooms (for a summary, see Bell & Burnaby, 1984, chapter 8; also see Bell, 1991). For the literacy learners, I often used the picture-words approach, following the principles of literacy instruction described by Wrigley (1993). We brought in actual cleaning supplies, materials, and equipment such as sponges, cleaning agents, mops, and vacuums, and pictures of furniture, pets, potted plants, and many other familiar objects.

Adverbial Expressions

Price is only X WHEN you clean the LR DR KT and Master BR

She will call us WHEN she wants more rooms cleaned.

AFTER a minute, the light will automatically go off.

WHEN you enter through the garage, the light *automatically* comes on.

WHEN locking the front door, please turn the key (see also participials)

Adjective Clauses

Read the note that is left by the customer before you start cleaning.

Conditionals

. . . to see if you will clean them.

If the customer is not home, call the office.

Turn off a/c in BR before vacuuming any of the upstairs; if you do not, you will *blow a fuse.*

Participials

When locking the front door, please turn the key ....

When dusting the *blinds*, put them back *as they were.*

When making the bed, make sure you put the *clips* back on the corners of the sheets.

Lock the back door *when finished.*

Directions, Location

N = north	L = left (R on S. Fairfax St.)
S = south	R = right

. . . becomes (name of St.) at dead end.

complex/subdivision	nearest intersection
apt. complex	
park; 2-hour parking	around to the right
on left side of N. X St.	
corners	edges
high places	

Names for Roads

BLVD = Boulevard	DR = Drive
CT = Court	PKWY = Parkway
HWY = Highway	L = Lane

Expressions of Time

biweekly	each time	on a regular basis

Furniture

baby changing table	piano bench
lighted china cabinet	sun porch
mantel	(table) legs

Continued on page 197

Instructions

Bring stepladder into house and clean top of refrigerator.

Thoroughly dust *baby changing table,* all shelves and legs.

Do not dust the *lighted china cabinet.*

Double-check the soap dishes in the shower for *soap film.*

Shake out dust from place mats after you dust *the lamps sitting on top of them.*

Please Windex *the tops* of the hanging mirrors *as well as the bottoms.*

Clean sliding glass door. Use *step stool* to *reach* top.

Take your time and make sure everything gets done.

Please turn the key *around to the right.*

Please clean *ends* of kitchen *counter.*

Don't use *air freshener*!

Do not *do* bathroom on second floor.

Special attention to the kitchen.

Do not walk on floors after *mopping*—you leave *footprints.*

Just vacuum and dust *sun porch.*

Take all trash.

Vacuum *then dust mop* the hardwood floors.

Turn off a/c in BR before vacuuming any of the upstairs; if you do not, you will *blow a fuse.*

Turn back on after vacuuming.

No beds (means do not do the beds)

Not cleaning the basement. (means We are not cleaning . . . .)

Put names of maids on quality cards; *she* likes to know who was there.

Method of Entry

MB   Admit Slip   Home

Miscellaneous

"No 1st entry fee; only paying $X"

"Customer owes us $XX from move-out cleaning in her old house; she will add it into 1st check."

Supplies and Materials

"her Endust"   "her Dow bathroom cleaner"   stepladder/stool

Verbs

adjust

blow (a fuse)

call ahead, put back (put them back)

check (also used as noun)

damp mop

double-lock (the front door)

Note. Italics indicate potentially troublesome vocabulary.

FIGURE 2. Sample Analysis of Field Sheets

One technique that worked well in our multilevel classroom was to use similar content with different activities for each level. For example, in a lesson on reporting damage, team leaders (and aspiring team leaders—those at high-intermediate and advanced levels) looked at broken items and practiced writing customer notes or made up customer-staff dialogues about the problem. Intermediate-level learners completed vocabulary-building activities and practiced the dialogues developed by the advanced-level participants. Beginning-level learners practiced telling a partner what had happened, and the literacy-level learners, in addition to participating in the dialogue practice, matched pictures with words or sentences describing what had happened.

We used other techniques to deal with classes made up of new and returning students. For example, if a lesson required sound knowledge of furniture vocabulary, new attendees received pictures with the furniture labeled whereas returning participants were encouraged to remember the words or look them up in previous lessons—or to ask the new participants to help them.

Generally, once we had chosen a topic and mapped out the content, methods logically followed. We also took learning styles and strategies into consideration. Beginning- and literacy-level learners liked word finds and crosswords, which tended to bore advanced-level or outgoing learners. Activities involving total physical response (TPR) were well received, so we used this method to teach the names of supplies and explain work processes (e.g., "I am putting the Ajax in the cleaning bucket. I am adding two sponges."). Housecleaning tasks such as vacuuming, taking out the trash, and changing the bed were also taught through TPR; a variation was letting the learners come up with these tasks in a language experience approach (LEA) story describing what they had done in a particular house. LEA was also effective for teaching workers how to tell their supervisors about work progress during the day or give a summary of the week's assignments. For advanced learners, we used these stories as openers, for example, to negotiate a less arduous cleaning schedule or a different sequence of cleaning jobs. The participants liked memorizing or reading dialogues, so we used this as a whole-group activity to begin and end every class. For example, if the day's topic was "Dealing With Damage," the dialogue involved a team leader telling a customer about some damage and what the company was going to do about it, or a team member telling one of the office staff (on the telephone) what had happened.

Other materials included partner-based cloze exercises and partner- or group-based information-gap exercises, problem-solving activities, and role plays. As noted above, often different groups of participants worked on different activities as the company assistants and I circulated throughout the room, helping as needed. The company supplied participants with picture dictionaries, which we used along with activities and audiocassette recordings. Advanced-level students made up exercises for testing one another, led activities for lower level students, and came up with a variety of self-directed learning activities. As the class progressed and participants' trust and degree of comfort grew, many would bring in notes they had tried to write based on an experience they had had at a home the previous week. Others would tell the group about a troublesome experience they had had. All of these student initiatives generated opportunities for productive learning activities. Review activities lent themselves well to games and cooperative learning approaches. Overt grammar instruction sometimes involved grammar-translation and other traditional methods.

Ideas from suggestopedia were useful for beginning-level learners, whose anxiety would sometimes become a problem.

Although we had to keep homework to a minimum because of work, family, and community demands on the learners' time, we assigned optional homework, which gave more practice to the students who were able to complete it. Participants were also encouraged to keep journals, make notes during the week about questions to raise in class, and perform many other small tasks that helped enhance learning.

The use of the learners' first language (L1) in class presented a dilemma: Many of the beginning-level learners needed to use their L1 as a learning support until they could negotiate meaning better, but the intermediate- and advanced-level learners would not benefit from hearing (or speaking) the L1 in class. By experimenting with different rules concerning learners' use of their L1, the class decided upon a flexible policy: Learners would try their best to use English but were allowed to use their L1 briefly when necessary. A particularly helpful way to stop overuse of the L1 was for me to speak to the offenders during a break to discover why they were relying on the L1, and then develop a solution jointly.

Evaluation and Feedback

Throughout the course, company managers received summaries of the performance milestones reached, for which they sometimes developed reinforcement mechanisms, either with me or independently. The organizational culture supported learning, and the company celebrated achievements, both large and small.

Testing to assess progress was challenging. One-on-one oral assessments were too time-consuming and expensive. The pretest and the first three attempts at unit tests showed good results for learners at the intermediate level and above but inconclusive results for the beginning- and literacy-level learners. For these learners, assessment involved ongoing monitoring of work completed in class, including audiocassettes of dialogues, TPR performance, and the like. As a means of ongoing evaluation, I had employees demonstrate their new skills both in and out of class. One method of evaluation included an error analysis of work completed by the participants in the first lesson (see Table 2). After a year, however, another round of program progress testing became necessary, and this time a second consultant was brought in to administer short oral assessments (no more than 5 minutes each) and a few written questions. Though not ideal, these methods were much more effective than written tests administered to the whole group.

◈ DISTINGUISHING FEATURES

Logistical Challenges

One distinctive feature of this program was the logistical challenges it faced and overcame. With open enrollment, staff scheduling, and employee absenteeism, class size could vary each week from 25 to 40 learners, including new employees, returning participants who attended irregularly, and those with perfect attendance. English proficiency ranged from literacy level through advanced level. At various points in the program, between two and five of the participants were enrolled in ongoing ESL programs at the local community college or a private language school; everyone else had little or no English training, and some had barely attended school

TABLE 2. SAMPLE ERROR ANALYSIS BASED ON WRITTEN EXERCISE

Possible Nature of Error	Examples From Stories	Teaching Points and Comments
Translation from native language (see below for more examples)	*no* for *not* "My son have two years." "Their family are six persons." "is very beautiful" "a very nice city, near is Hollywood."	Show how these items translate into English
Subject pronoun drop	"I love [X] because is beautiful." "I love my country is to small."	Sentence structure
Verb tense	All tenses (esp. perfect) and forms "I live there 3 years ago." "There are lived in Alex." "I come to U.S. six years ago because . . . ." "I live in U.S. for a long time." "I live in the U.S. for a long time ago." "I liven an Dapartments la union. "Now I'm living Alex. one year ago." "My family living in Mexico."	Verbs • conjugation • forms • in context • shifting • with other words (e.g., *ago*)
Subject-verb agreement	"My son have . . . ." ". . . city have . . . ."	Subject-verb agreement How L1 expressions translate
Singular/plural	"three son " "1 children " "many place tourist"	English singular/plural rules
Expressions of quantity	"I love my country is to small" "and have too much" river. ". . . too much people" ". . . too much hot" ". . . too hot and the sun shines every day it's beautiful." "I love too much my parents."	Expressions of quantity
Infinitives after certain verbs	"I need help my mother."	Gerunds and infinitives after certain verbs
Articles	Dropping *a* Using *the* for *a* "*una* sister-in-law"	Low priority, given the other problems
Prepositions	"I live Alexandria" "a lot trees" "many place tourist"	Prepositions • dropping • using wrong preposition Low priority unless it affects meaning

Continued on page 201

TABLE 2 (*continued*). SAMPLE ERROR ANALYSIS BASED ON WRITTEN EXERCISE

Possible Nature of Error	Examples From Stories	Teaching Points and Comments
Adjective clauses and phrases	"I lived in L.A. where living to many actors and singers very famous." "there are a lot of trees very nice." "The city I was lived is very hot . . . .". "and my fadher it is good fisherman"	Adjective clauses and phrases
Adverbials	Many errors around *for, ago,* not using *since,* and trouble using the perfect tense (see Verb Tenses) "I live in the U.S. for a long time ago." "I came about two years." (*ago*) See errors under Verb Tenses "I love too much my parents." ". . . and never it is cold."	Adverb clauses and phrases Adverb placement
Spelling	Most common: *is* for *it's; is* for *it* *to* for *too* *an* for *and* *mather* for *mother* *tree* for *three* Many variations for *daughter* Others: *lots* for *last* *de* for *the* *berry* for *very*	Spelling rules Phonics
Linking verb *be*	"I very happy because . . ."	Linking verb *be*
There is/are	"My country is small, but is to many people." "Are many mountains."	*There is/are*
Miscellaneous	At the end of an otherwise advanced-level paper, after listing the ages of her children, participant wrote: "I brind my little son tree mouth"	Inconsistency is interesting; check on participant's performance during class.
	Using *I have* in the following, which were four consecutive lines from one paper (note Lines 3 and 4): "I have 2 son and 3 sister. "My contry is El Salvador." "I have beautiful beach. "I have many place tourist."	2nd example—check on ? Check on understanding of *I have* versus *It has*.
	"My modher is tailor in my country." She is Household Service in my family and my fadher it is good fisherman beautiful in my country."	? Use of *beautiful* in last line.

either in the United States or in their native country. We addressed this challenge by using the methods appropriate to multilevel classrooms described in the Instructional Strategies and Resources section and taking advantage of the company's willingness to provide on-the-job reinforcement of the learning.

Another logistical challenge was classroom space. As is typical, the company site was set up to conduct business rather than classes. There was not enough table space for all the participants, so more than half of the learners had to write on their laps while holding their dictionary and class book. The room had a single flip chart on an easel but no whiteboards or blackboards, although it had more than enough electrical outlets, and a videocassette player and monitor parked in a corner.

Company Support

The company showed a high level of commitment to the program and instinctively understood how to reinforce the learning on the job. The company's commitment permeated every aspect of the business and the training, which helped the program achieve success. Examples of this support include the following:

- The company's owner immediately institutionalized the program by incorporating it into weekly staff meetings and informing everyone in the company about the program, its objectives, and the importance of supporting the participants' learning.

- Company staff served as teacher's aides to help deal with the large, multilevel group.

- The company informed customers about the class and encouraged them to give feedback on the cleaning staff's progress in English.

- The owner established a feedback loop within the company that included the staff, management, participants, and me. We worked together to ensure the program's success.

- The company promised participants that lack of progress in class would not result in negative employment decisions; however, the English class served as developmental training to enhance team members' ability to progress. Team members who improved their English proficiency, for example, were promoted to team leaders.

- The company made English class a priority; rarely was class interrupted for business reasons, and in $3^{1}/_{2}$ years it was canceled only twice for work crises.

- The owner extended the company's positive work atmosphere to the English program, making conditions safe for participants to try out their new English and helping the cleaning staff apply their new skills to their work more quickly.

The management and office staff had an unusually acute sense of what company resources would be useful to the ESL class. Often in workplace ESL programs, consultants who do not ask the right questions about available resources might miss out on having access those resources. (Even when the right questions are asked, the company personnel often cannot relate the request to the resources available throughout the company.) In this program, the company staff's understanding of what would be helpful, combined with frequent contact with and discussions about the program, enabled a steady supply of company materials that helped bring authenticity into the classroom.

Another means of support was the owner's leadership skills. He permitted and even encouraged a great deal of experimentation, flexibility, and risk taking. Dealing with a large, multilevel, multineed class with constantly changing enrollment required flexibility, adaptability, and most important, experimentation. Because the owner fostered a sense of entrepreneurial freedom throughout the company, the program achieved its objectives despite sometimes almost impossible odds.

◈ PRACTICAL IDEAS

Even ESP programs that appear to be incompatible with certain business environments can be successful. The ESP program described here overcame obstacles that are inherent to providing English training in a busy, small business environment by being flexible in its perspective, focusing on the positive aspects of the situation, and being willing to take risks.

View Matters From the Other's Perspective

The business manager and I each viewed needs and realities from the other's perspective. In the business discussed in this chapter, demand for service was high and growing, employee recruitment was difficult, staff shortages occurred, and workload was peaking. In this environment, conditions that would have been ideal from my perspective—multiple classes per week, small classes, static enrollment, and consistent attendance—were impossible from the perspective of the business owner, as they would result in lost revenue, disrupted schedules, and possible loss of customers.

Once ESP practitioners recognize the constraints placed on a program because of the business environment in which it occurs, they must ask themselves whether or not these constraints will affect a program's results to the point that it is not feasible. The practitioner must discuss with the owner the constraints and their likely effect on the program. For example, in this program, I discussed with the manager the potential problems of holding only one very large, multilevel class held just once a week for less than half a morning. It is important for the business owner to understand the trade-offs: Although the program's structure disrupts the work schedule, the result may be less learning, and the business may not receive the results from the training that it had hoped for.

Make the Most of the Situation

If the alternative to a less-than-ideal situation is no training whatsoever, then the practitioner may decide to be more flexible about the conditions under which the program operates. The situation may seem hopeless when 21–40 students attend a multilevel class in a given week, some students have attended consistently for 11 weeks and others are attending for the first time, some have materials and some do not, and some are literate and some are not. Yet these conditions do not render the situation hopeless. The practitioner can meet challenges by carefully selecting teaching methods and materials, maintaining rapport with the learners and managers, and adjusting lesson plans based on actual attendance.

◈ CONCLUSION

Despite the many challenges discussed in this case study, this ESP program helped employees improve their knowledge and use of English on the job in significant ways. By understanding each other's challenges, working within the constraints, and accepting trade-offs, business owners and language specialists can work together to create successful ESP programs.

◈ CONTRIBUTOR

Patricia Noden, who owns and operates Step-Wise Training & Consulting, has been conducting ESL programs since 1986. She holds a BS in public management, an MPA, an MA in linguistics, and a TESL certificate from George Mason University. She is a member of TESOL, the National Association of Female Executives, and the American Society of Training and Development.

References

Alsberg, J. (in press). Effecting change in pronunciation: Teaching ITAs to teach themselves. In *Proceedings of the Sixth National Conference on the Education and Employment of Graduate Teaching Assistants*. Stillwater, OK: New Forums Press.

Analyzing cases. (1992). Unpublished manuscript, Columbia University, School of Business and Management of Organizations Division, New York.

Atlas, J. (1999, July 19). The million-dollar diploma. *The New Yorker*, 42–51.

Auerbach, E. R. (1992). *Making meaning, making change: Participatory curriculum development for adult ESL literacy*. McHenry, IL: Delta Systems.

Bailey, K. M. (1984). The "foreign TA problem." In K. M. Bailey, F. Pialorsi, & J. Zukowski/Faust (Eds.), *Foreign teaching assistants in U.S. universities* (pp. 4–6). Washington, DC: National Association of Foreign Student Advisors.

Baker, M. (1979). *English for nautical students*. Glasgow, Scotland: Brown, Son & Ferguson.

Barton, L. (1993). *Crisis in organizations: Managing and communicating in the heat of chaos*. Cincinnati, OH: South-Western.

Basturkmen, H. (1999). Discourse in MBA seminars: Towards a description for pedagogical purposes. *English for Specific Purposes, 18*, 63–80.

Beetham, J. (1999, March/April). Business English in Europe: First among equals? *American Language Review*, 66.

Beigbeder, F. (1997). *Nuevo diccionario politécnico de las lenguas española e inglesa*. Madrid, Spain: Editorial Díaz de Santos.

Bell, J. (1991). *Teaching multi-level classes in ESL*. San Diego, CA: Dominie Press.

Bell, J., & Burnaby, B. (1984). *A handbook for ESL literacy*. Toronto, Canada: Ontario Institute for Studies in Education Press.

Bhatia, V. K. (1993). *Analysing genre: Language use in professional settings*. London: Longman.

Bhatia, V. K. (2000, March). *Integrating discursive competence and professional practice: A new challenge for ESP*. Paper presented at the 34th Annual TESOL Convention, Vancouver, BC, Canada.

Black, H. C. (1991). *Black's law dictionary* (6th ed.). St. Paul, MN: West.

Blackman, I. (1993). *English language skill and diversity in nursing practice: A vocational and educational response to the presence of non-English speaking background nurses in Australia*. Unpublished master's thesis, University of South Australia, Adelaide.

Blakey, T. N. (1987). *English for maritime studies*. London: Prentice-Hall International.

Blicq, R. (1993). *Technically—write! Communicating in a technological era* (4th ed.). Englewood Cliffs, NJ: Prentice Hall.

Boyd, F. (1991). Business English and the case method: A reassessment. *TESOL Quarterly, 25*, 729–734.

Boyd, F. (1994). *Making business decisions: Real cases from real companies.* White Plains, NY: Longman.

Boyle, E. R. (1995, November). *Negotiated syllabus in EAP business English classes.* Paper presented at the Annual Meeting of the Japanese Association of Language Teachers, Nagoya, Japan. (ERIC Document Reproduction Service No. ED404860)

Boyter-Escalona, M. (1998). *Final report of the Workplace Literacy Partnership program.* Washington, DC: U.S. Department of Education, National Workplace Literacy Program. (ERIC Document Reproduction Service No. ED425285)

Canseco, G., & Byrd, P. (1989). Writing required in graduate courses in business administration. *TESOL Quarterly, 23,* 305–316.

Carnevale, A. P., Gainer, L. J., & Meltzer, A. S. (1988). *Workplace basics: The skills employers want.* Washington, DC: American Society for Training and Development and U.S. Department of Labor.

Chisman, F. P. (1992). *The missing link: Workplace education in small businesses.* Washington, DC: Southport Institute for Policy Analysis.

Cohen, A., Glassman, H., Rosenbaum-Cohen, P. R., Ferradas J., & Fine, J. (1979). Reading for specialized purposes: Discourse analysis and the use of student informants. *TESOL Quarterly, 13,* 157–167.

Cole, S. A., & Bird, J. (2000). *The medical interview: The three-function approach* (2nd ed.). St. Louis: Mosby.

Connor, U. (1996). *Contrastive rhetoric: Cross-cultural aspects of second-language writing.* Cambridge: Cambridge University Press.

Conti v. ASPCA, 77 Misc.2d 61, 353 N.Y.S.2d 288 (N.Y. Civ. Ct. 1974).

Danet, B. (1985). Legal discourse. In T. Van Dijk (Ed.), *Handbook of discourse analysis* (Vol. 1, pp. 273–291). London: Academic Press.

Davidson, W. (1994). *Business writing: What works, what won't.* New York: St. Martin's Press.

DiStefano, J. (1974). *International Bank of Malaysia.* London, Ontario, Canada: University of Western Ontario, School of Business Administration.

Drucker, P. F. (1992, September/October). The new society of organizations. *Harvard Business Review, 7*(5), 95–104.

Dudley-Evans, T. (1988). Recent developments in ESP: The trend to greater specialisation. In M. L. Tickoo (Ed.), *ESP: State of the art* (pp. 27–32). Singapore: SEAMEO Regional Language Centre.

Dudley-Evans, T., & St. John, M. J. (1998). *Developments in English for specific purposes: A multidisciplinary approach.* Cambridge: Cambridge University Press.

Dunn, M. R., Miller, R. S., & Richter, T. H. (1998). Graduate medical education 1997–1998. *Journal of the American Medical Association, 280,* 836–845.

Educational Testing Service. (1996). *The SPEAK rater training kit.* Princeton, NJ: Author.

Eggly, S., Afonso, N., Rojas, G., Baker, M., Cardozo, L., & Robertson, R. S. (1997). An assessment of residents' competence in the delivery of bad news to patients. *Academic Medicine, 72,* 397–399.

Eyres, D. J. (1990). *Ship construction.* Oxford: Heinemann Newnes.

Feak, C. B., Reinhart, S. M., & Sinsheimer, A. (2000). A preliminary analysis of law review notes. *English for Specific Purposes, 19,* 197–220.

Fisher, R., & Ury, W. (1992). *Getting to yes: Negotiating agreement without giving in* (2nd ed.). New York: Penguin Books.

Fredrickson, K. M. (1995). *American and Swedish written legal discourse: The case of court documents.* Unpublished doctoral dissertation, The University of Michigan, Ann Arbor.

Fredrickson, K. M. (1998). Languages of the law: A course in English for legal studies. *The Language Teacher, 22*(11), 23–25, 39.

Freire, P. (1973). *Education for critical consciousness.* New York: Seabury Press.

Floyd, P., & Carrell, P. L. (1994). Effects on ESL reading of teaching cultural content schemata. In A. H. Cumming (Ed.), *Bilingual performance in reading and writing* (pp. 309–329). Ann Arbor, MI: John Benjamins.

Galanti, G. (1997). *Caring for patients from different cultures: Case studies from American hospitals* (2nd ed.). Philadelphia: University of Pennsylvania Press.

Garcia, P., & Sharma, S. (1995). *Workplace communication and computer-assisted learning.* Chicago: Northeastern Illinois University. (ERIC Document Reproduction Service No. ED425289)

Gardner, H. (1993). *Frames of mind: The theory of multiple intelligences.* New York: Basic Books.

Graham, J. L. (1985). The influence of culture on the process of business negotiations: An exploratory study. *Journal of International Business Studies, 16,* 79–94.

Hadfield, J. (1984). *Communication games.* Walton on Thames, England: Nelson Harrap.

Hahn, L., & Dickerson, W. (1999a). *Speechcraft: Discourse pronunciation for advanced learners.* Ann Arbor: University of Michigan Press.

Hahn, L., & Dickerson, W. (1999b). *Speechcraft: Workbook for international TA discourse.* Ann Arbor: University of Michigan Press.

Harris, S. (1992). Reaching out in legal education: Will EALP be there? *English for Specific Purposes, 11,* 19–32.

Hinkin, T. R. (1995). *Cases in hospitality management: A critical incident approach.* New York: Wiley.

Holec, H. (1981). *Autonomy and foreign language learning.* Oxford: Oxford University Press.

Howe, P. M. (1990). The problem of the problem question in English for academic purposes. *English for Specific Purposes, 9,* 215–236.

Hussin, V. (1991). *Promoting positive clinical learning experiences for overseas qualified nurses of non-English-speaking background.* Paper presented at the Preceptor Workshop, School of Nursing, The Flinders University of South Australia, Adelaide.

Hutchinson, T., & Waters, A. (1987). *English for specific purposes: A learning-centred approach.* Cambridge: Cambridge University Press.

Iacobelli, C. L. (1993, March). *A business English curriculum in an academic setting* [Abstract]. Paper presented at the Annual Eastern Michigan University Conference on Languages and Communication for World Business and the Professions, Ypsilanti, MI. (ERIC Document Reproduction Service No. ED370384)

Inglehart, J. K. (1996). Health policy report: The quandary over graduates of the foreign medical schools in the United States. *New England Journal of Medicine, 334,* 1679–1683.

Ingram, D., & Wylie, D. (1995). *Australian second language proficiency rating.* Queensland, Australia: Griffith University, Centre for Applied Linguistics and Language.

Inman, M. (1985). Language and cross-cultural training in American multinational corporations. *The Modern Language Journal, 69,* 247–255.

InterDis Research Group. (n.d.). *Curso multimedia de inglés naval* [Computer software]. Unpublished.

Jenkins, S., & Hinds, J. (1987). Business letter writing: English, French, and Japanese. *TESOL Quarterly, 22,* 327–345.

Johns, A. M., & Dudley-Evans, T. (1991). English for specific purposes: International in scope, specific in purpose. *TESOL Quarterly, 25,* 297–314.

Johns, A. M., & Dudley-Evans, T. (1993). English for specific purposes: International in scope, specific in purpose. In S. Silberstein (Ed.), *State-of-the-art TESOL Essays: Celebrating 25 years of the discipline* (pp. 115–132). Alexandria, VA: TESOL.

Johnston, W. B., & Packer, A. H. (1987). *Workforce 2000: Work and workers for the twenty-first century.* Indianapolis, IN: Hudson Institute.

Jordan, M. (1984). *Rhetoric of everyday English texts.* London: Allen & Unwin.

Kaufman, D., & Brooks, J. G. (1996). Interdisciplinary collaboration in teacher education: A constructivist approach. *TESOL Quarterly, 30,* 231–251.

Kavanaugh, K. (1999, April). Teaching the language of work. *Training & Development Magazine, 53*(4), 14–16.

Kemp, J. F. (1997). *Ship construction sketches and notes*. Oxford: Butterworth-Heinemann.

Kintsch, W. (1994). The psychology of discourse processing. In M. A. Gernsbacher (Ed.), *Handbook of psycholinguistics* (pp. 721–739). San Diego, CA: Academy Press.

Kintsch, W. (1988). The role of knowledge in discourse comprehension: A construction-integration model. *Psychological Review, 95,* 163–182.

Krahnke, K. (1987). *Approaches to syllabus design for foreign language teaching*. Englewood Cliffs, NJ: Prentice Hall.

Krashen, S. (1982). *Principles and practice in second language acquisition*. Oxford: Pergamon Press.

Krashen, S. (1983). Applications of psycholinguistic research to the classroom. In M. Long & J. Richards (Eds.), *Methodology in TESOL: A book of readings* (pp. 33–44). Rowley, MA: Newbury House.

Larmer, B. (1999, July 12). Latin U.S.A. *Newsweek,* p. 48.

Lazear, D. (1991). *Seven ways of knowing: Teaching for multiple intelligences*. Palatine, IL: Skylight.

Literacy and Education Research Network, New South Wales Department of School Education. (1990). *A genre-based approach to teaching writing, Book 2*. Annandale, Australia: Common Ground.

London Chamber of Commerce and Industry, Examinations Board. (1995). *English for the tourism industry: Extended syllabus*. Sidcup, England: Author.

A look at English education and corporate training programs implemented by foreign-owned companies in Japan. (1999, February). *TOEIC Newsletter,* 2–12.

López, E., Spiegelberg, J. M., & Carrillo, F. (1998). *Inglés técnico naval* (3rd ed.). Cádiz, Spain: Servicio de Publicaciones de la Universidad de Cádiz.

Louhiala-Salminen, L. (1996). The business communication classroom vs. reality: What should we teach today? *English for Specific Purposes, 15,* 37–51.

Lundeberg, M. A. (1987). Metacognitive aspects of reading comprehension: Studying understanding in legal case analysis. *Reading Research Quarterly, 22,* 407–432.

Martin, S. B., & Garcia, P. (1997). *Administering the General Work-Based Assessment*. Chicago: Northeastern Illinois University. (ERIC Document Reproduction Service No. ED413793)

Malagón Ortuondo, J. M. (1998). *Diccionario náutico. Inglés-español. Español-inglés*. Madrid, Spain: Editorial Paraninfo.

McBurney, N. (1996). *Tourism* (Professional Reading Skills Series). Hemel Hempstead, England: Prentice Hall.

McGroarty, M. (1993). Second language instruction in the workplace. In W. Grabe (Ed.), *Annual review of applied linguistics* (pp. 86–108). New York: Cambridge University Press.

Micheau, C., & Billmyer, K. (1987). Discourse strategies for foreign business students: Preliminary research findings. *English for Specific Purposes, 16,* 87–97.

The Miller Brewing Company. (1990). *Brewing beer: An overview* (Profit Through Knowledge series) [videotape]. Milwaukee, WI: Author.

Milton, J., & Jacobs, G. (1995). Teaching technical vocabulary with multimedia resources. In A. Gimeno (Ed.), *Proceedings of EUROCALL '95* (pp. 321–327). Valencia, Spain: Servicio de Publicaciones de la Universidad Politécnica de Valencia.

Monoson, P. K., & Thomas, C. F. (1993). Oral English proficiency policies for faculty in U.S. higher education. *Review of Higher Education, 16,* 127–140.

Morrow, P. R. (1995). English in a Japanese company: The case of Toshiba. *World Englishes, 14,* 87–98.

Munby, J. (1978). *Communicative syllabus design: A sociolinguistic model for defining the content of purpose-specific language programmes*. Cambridge: Cambridge University Press.

Munter, M. (1999). *Guide to managerial communication* (5th ed.). Englewood Cliffs, NJ: Prentice Hall.

Murphy, B. & Pascoe, A. (1996). *Using the Internet on a business English course* [Abstract]. (ERIC Document Reproduction Service No. ED403743)

Nadler, L. (1988). *Designing training programs.* New York: Addison-Wesley.

Neu, J. (1986). American English business negotiations: Training for non-native speakers. *English for Specific Purposes, 5,* 41–57.

Normal beer and nonalcoholic substitutes. (1915, October). *American Brewer.*

Nunan, D. (1985). *Language teaching course design: Trends and issues.* Adelaide, Australia: National Curriculum Resource Centre.

Nunan, D. (1992). *Research methods in language learning.* Cambridge: Cambridge University Press.

Nunan, D. (1995). Introduction. In *Atlas* (Teacher's extended ed., pp. xi–xii). Boston: Heinle & Heinle.

Nunan, D. (1997, September). Listening in language learning. *The Language Teacher Online.* Retrieved from http://langue.hyper.chubu.ac.jp/jalt/pub/tlt/97/sep/nunan.html.

O'Barr, W. M. (1982). *Linguistic evidence: Language power and strategy in the courtroom.* New York: Academic Press.

Open doors 1997–98: Report on international educational exchange. (1998). New York: Institute of International Education.

Orsi, L., & Orsi, P. (1990). *Brewing beer/Proceso de elaboración de la cerveza.* Quilmes, Argentina: Cervecería y Maltería Quilmes.

Papajohn, D. (1997, November). *The standard setting process for the new Test of Spoken English: A university case study.* Paper presented at the Sixth National Conference on the Education and Employment of Graduate Teaching Assistants, Minneapolis, MN.

Papajohn, D. (1998). *Toward speaking excellence: The Michigan guide to maximizing your performance on the TSE test and SPEAK test.* Ann Arbor: The University of Michigan Press.

Paradiso, J., & Pawlowski, K. (1993). *From the grass: An interview with Dr. Krzysztof Pawlowski, Rector of WSB-NLU in Nony Sacz, Poland* [Abstract]. (ERIC Document Reproduction Service No. ED368434)

Pearson, J. (1998). *Terms in context.* Amsterdam: John Benjamins.

Perreault, W. D., & McCarthy, E. J. (1999). *Basic marketing: A global-managerial approach* (12th ed.). Boston: Irwin McGraw-Hill.

Phillips, J. J. (1997). *Return on investment in training and performance improvement programs.* Houston, TX: Gulf.

Pohl, A. (1996). *Test your business English: Hotel and catering.* London: Penguin Books.

Poor, E. (1992). *The executive writer: A guide to managing words, ideas, and people.* New York: Grove Weidenfeld.

Pursey, H. J. (1998). *Merchant ship construction.* Glasgow, Scotland: Brown, Son & Ferguson.

Richards, K. (1989). Pride and prejudice: The relationship between ESP and training. *English for Specific Purposes, 8,* 202–222.

Rideout, C. (1991). Research and writing about legal writing. *Legal Writing, 1,* v–ix.

Rivers, W., & Melvin, B. S. (1977). If only I could remember it all! Facts and fiction about memory in language learning. In M. Burt, H. Dulay, & M. Finocchiaro (Eds.), *Viewpoints on English as a second language* (pp. 162–171). New York: Regents.

Robinson, D. G., & Robinson, J. C. (1989). *Training for impact: How to link training to business needs and measure the results.* San Francisco: Jossey-Bass.

Robinson, D. G., & Robinson, J. C. (1995). *Performance consulting: Moving beyond training.* San Francisco: Berrett-Koehler.

Robinson, D. G., & Robinson, J. C. (1998). *Moving from training to performance: A practical guidebook.* San Francisco: Berrett-Koehler.

Robinson, P. (1991). *ESP today: A practitioner's guide*. New York: Prentice Hall.

Rosenblatt, L. M. (1994). The transactional theory of reading and writing. In R. Ruddell, M. R. Ruddell, & H. Singers (Eds.), *Theoretical models and processes of reading* (4th ed., pp. 1057–1092). Newark, DE: International Reading Association.

Rothwell, W. J., & Brandenburg, D. C. (1990). *The workplace literacy primer*. Amherst, MA: Human Resource Development Press.

Sarmiento, A. R., & Kay, A. (1990). *Worker-centered learning: A union guide to workplace literacy*. Washington, DC: AFL-CIO Human Resources Development Institute.

Schleppegrell, M., & Royster, L. (1990). Business English: An international survey. *English for Specific Purposes, 9,* 3–16.

Selinker, L. (1988). *Papers in interlanguage* (Occasional Paper No. 44). Singapore: SEAMEO Regional Language Centre. (ERIC Document Reproduction Service No. ED321549)

Shapiro, N. (1994). *Chalk talk*. Berkeley, CA: Command Performance Language Institute.

Shapo, H. S., Walter, M. R., & Fajans, E. (1991). *Writing and analysis in the law*. Westbury, NY: The Foundation Press.

Silverman, J., Kurtz, S., & Draper, J. (1999). *Skills for communicating with patients*. Oxford: Radcliff Medical Press.

Sinhaneti, K. (1994). *ESP courses at tertiary level: A reflection of the Thai business community* [Abstract]. (ERIC Document Reproduction Service No. ED377724)

Smith, F. (1985). *Reading without nonsense*. New York: Teachers College Press.

Smith, J., Meyers, C., & Burkhalter, A. (1992). *Communicate: Strategies for international teaching assistants*. Englewood, NJ: Regents/Prentice Hall.

Smith, R., Byrd, P., Nelson, G., Barrett, R., & Constantinides, C. (1992). *Crossing pedagogical oceans: International teaching assistants in U.S. undergraduate education* (ASHE-ERIC Higher Education Report No. 8). Washington, DC: The George Washington University, School of Education & Human Development.

Soifer, R., Irwin, M. E., Crumrine, B. M., Honzaki, E., Simmons, B. K., & Young, D. L. (1990). *The complete theory-to-practice handbook of adult literacy: Curriculum design and teaching approaches*. New York: Teachers College Press.

Spada, N. (1990). Observing classroom behaviors and learning outcomes in different second language programs. In J. C. Richards & D. Nunan (Eds.), *Second language teacher education* (pp. 293–310). Cambridge: Cambridge University Press.

St. John, M. (1996). Business is booming: Business English in the 1990s. *English for Specific Purposes, 15,* 3–18.

Stake, R. E. (1995). *The art of case study research*. Thousand Oaks, CA: Sage.

State Tourism Training Agency. (1997). *Tourism and travel in Ireland* (2nd ed.). Dublin, Ireland: Gill & McMillan.

Strevens, P. (1988). ESP after twenty years: A re-appraisal. In M. L. Tickoo (Ed.), *ESP: State of the art* (pp. 1–13). Singapore: SEAMEO Regional Language Centre.

Suárez Gil, L. (1983). *Diccionario técnico marítimo*. Madrid, Spain: Editorial Alhambra.

Svendsen, C., & Krebs, K. (1984). Identifying English for the job: Examples from health care occupations. *ESP Journal, 3,* 153–164.

Swales, J. (1986). ESP comes of age? 21 years after "Some Measurable Characteristics of Scientific Prose." In J. Swales, *English for specifiable purposes* (Occasional Paper No. 42, pp. 1–11). Singapore: SEAMEO Regional Language Centre.

Swales, J. (1990). *Genre analysis: English in academic and research settings*. Cambridge: Cambridge University Press.

Syratt, G. (1992). *Manual of travel agency practice*. Oxford: Butterworth-Heinemann.

Syratt, G. (1995). *Manual of travel agency practice* (2nd ed.). Oxford: Butterworth-Heinemann.

Tackling a big job. (1915, October). *American Brewer.*

Thomas, R. J. (1991). *Job-related language training for limited English proficient employees: A handbook for program developers*. Washington, DC: Development Assistance Corp.

Thong, I. (1995, April). *Language and institutional capacity building in Cambodia: A case study of the faculty of business in Phnom Penh*. Paper presented at the International Conference on Language in Development, Denpasar, Indonesia. (ERIC Document Reproduction Service No. ED389221)

Tomizawa, S. (1991, April). *Designing an intensive English program for business people: Curriculum and courses*. Paper presented at the Annual Eastern Michigan University Conference on Languages and Communications for World Business and the Professions, Ypsilanti, MI. (ERIC Document Reproduction Service No. ED344485)

Tupper, E. (1996). *Introduction to naval architecture*. Oxford: Butterworth-Heinemann.

University of Cambridge Local Examinations Syndicate. (1998). *ALTE handbook*. Cambridge: Author.

U.S. Bureau of the Census. (1998). *State and metropolitan area data book 1997–98* (5th ed.). Washington, DC: U.S. Government Printing Office.

U.S. Department of Education & U.S. Department of Labor. (1988). *The bottom line: Basic skills in the workplace*. Washington, DC: U.S. Government Printing Office.

Uvin, J. (1996). Designing workplace ESOL courses for Chinese health-care workers at a Boston nursing home. In K. Graves (Ed.), *Teachers as course developers* (pp. 39–62). New York: Cambridge University Press.

Virginia Employment Commission. (n.d.). *Industry and occupational employment projections: 1998–2008*. Richmond, VA: Author. Retrieved July 14, 2000, from http://www.vec.state .va.us/pdf/novaproj.pdf.

Vygotsky, L. S. (1980). *Mind in society: The development of higher psychological processes* (M. Cole, V. John-Steiner, S. Scribner, & E. Souberman, Eds.). Cambridge, MA: Harvard University Press.

West, R., & Walsh, G. (1993). *ESU framework: Performance scales for English language examinations* (2nd ed.). London: Longman.

Widdowson, H. (1979). *Explorations in applied linguistics*. Oxford: Oxford University Press.

Williams, M. (1988). Language taught for meetings and language used in meetings: Is there anything in common? *Applied Linguistics, 9,* 45–58.

Worker Education Program. (1994). *General workplace curriculum guide: English as second language for the workplace*. Chicago: Northeastern Illinois University. (ERIC Document Reproduction Service No. ED392310)

Wright, A. (1984). *1000 pictures for teachers to copy*. Reading, MA: Addison-Wesley.

Wrigley, H. S. (1993). *Innovative programs and promising practices in adult ESL literacy* (ERIC Digest EDO-LE-93-07). Washington, DC: National Clearinghouse on Literacy Education. (ERIC Document Reproduction Service No. ED358748)

Yin, K. M., & Wong, I. (1990). A course in business communication for accountants. *English for Specific Purposes, 9,* 253–263.

Yin, R. K. (1994). *Case study research: Design and methods* (2nd ed.). Thousand Oaks, CA: Sage.

Yogman, J., & Kaylani, C. T. (1996). ESP program design for mixed level students. *English for Specific Purposes, 15,* 311–324.

Index

Page numbers followed by *f* and *t* indicate figures and tables, respectively.

Also Available From TESOL

Academic Writing Programs
Ilona Leki, Editor

Action Research
Julian Edge, Editor

Bilingual Education
Donna Christian and Fred Genesee, Editors

CALL Environments:
Research, Practice, and Critical Issues
Joy Egbert and Elizabeth Hanson-Smith, Editors

Community Partnerships
Elsa Auerbach, Editor

Distance-Learning Programs
Lynn E. Henrichsen, Editor

Implementing the ESL Standards for Pre-K–12 Students Through Teacher Education
Marguerite Ann Snow, Editor

Integrating the ESL Standards Into Classroom Practice: Grades Pre-K–2
Betty Ansin Smallwood, Editor

Integrating the ESL Standards Into Classroom Practice: Grades 3–5
Katharine Davies Samway, Editor

Integrating the ESL Standards Into Classroom Practice: Grades 6–8
Suzanne Irujo, Editor

Integrating the ESL Standards Into Classroom Practice: Grades 9–12
Barbara Agor, Editor

Intensive English Programs in Postsecondary Settings
Nicholas Dimmitt and Maria Dantos-Whitney, Editors

Internet for English Teaching
Mark Warschauer, Heidi Shetzer, and Christine Meloni

Journal Writing
Jill Burton and Michael Carroll, Editors

Managing ESL Programs in Rural and Small Urban Schools
Barney Bérubé

Reading and Writing in More Than One Language:
Lessons for Teachers
Elizabeth Franklin, Editor

Teacher Education
Karen E. Johnson, Editor

Teaching in Action: Case Studies From Second Language Classrooms
Jack C. Richards, Editor

Technology-Enhanced Learning Environments
Elizabeth Hanson-Smith, Editor

For more information, contact
Teachers of English to Speakers of Other Languages, Inc.
700 South Washington Street, Suite 200
Alexandria, Virginia 22314 USA
Tel 703-836-0774 • Fax 703-836-6447 • publications@tesol.org •
http://www.tesol.org/